SpringerBriefs in Statistics

For further volumes:
http://www.springer.com/series/8921

George A. F. Seber

Statistical Models for
Proportions and Probabilities

 Springer

George A. F. Seber
Department of Statistics
The University of Auckland
Auckland
New Zealand

ISSN 2191-544X   ISSN 2191-5458   (electronic)
ISBN 978-3-642-39040-1 ISBN 978-3-642-39041-8   (eBook)
DOI 10.1007/978-3-642-39041-8
Springer Heidelberg New York Dordrecht London

Library of Congress Control Number: 2013942014

Printed on acid-free paper

Springer is part of Springer Science+Business Media (www.springer.com)

Preface

Most elementary statistics books discuss inference for proportions and probabilities, and the primary readership for this monograph is the student of statistics, either at an advanced undergraduate or graduate level. As some of the recommended so-called "large-sample" rules in textbooks have been found to be inappropriate, this monograph endeavors to provide more up-to-date information on these topics. I have also included a number of related topics not generally found in textbooks. The emphasis is on model building and the estimation of parameters from the models.

It is assumed that the reader has a background in statistical theory and inference and is familiar with standard univariate and multivariate distributions, including conditional distributions. This monograph may also be helpful for the statistics practitioner who is involved with statistical consulting in this area, particularly with regard to inference for one and two proportions or probabilities.

Chapter 1 looks at the difference between a proportion and probability. It focuses on a proportion leading to the Hypergeometric model and its Binomial approximation, along with inference for the proportion. Inverse sampling is also considered. Chapter 2 focuses on estimating a probability and considers the Binomial distribution in detail as well as inverse sampling. Exact and approximate inferences for a probability are considered. In Chap. 3, the main focus is on comparing two proportions or two probabilities and related quantities such as the relative risk and the odds ratio from the same or different populations using the Multi-hypergeometric or Multinomial distributions. Simultaneous confidence intervals for several parameters are also considered. The Multinomial distribution is the basis for a number of hypothesis and goodness of fit tests, and these are discussed in Chap. 4 with particular attention given to 2×2 tables and matched data. In Chap. 5, we look briefly at two logarithmic models for discrete data, namely the log linear and the logistic models.

I would like to thank two reviewers for their very helpful comments on a previous draft.

Auckland, New Zealand, June 2012   George A. F. Seber

Contents

Chapter 1
Single Proportion

Abstract This chapter focusses on the problem of estimating a population proportion using random sampling with or without replacement, or inverse sampling.
Exact and approximate confidence intervals are discussed using the Hypergeometric
distribution. Applications to capture-recapture models are given.

Keywords Proportion · Hypergeometric distribution · Simple random sample ·
Sampling fraction · Negative-Hypergeometric distribution · Binomial distribution ·
Confidence intervals for a proportion · Single capture-recapture model

1.1 Distribution Theory

Since this monograph is about modelling proportions and probabilities, I want to
begin by comparing the two concepts, proportion and probability, as these two ideas
are sometimes confused. If we have a population of N people and M are male, then
the proportion of males in the population is $p = M/N$. Suppose we now carry out
a random experiment and choose a person at random from the population. What we
mean by this is that we choose a person in such a way that every person is equally
likely to be chosen. If the population is small we could achieve this by putting the
names of everyone in a container, shuffling the names by rotating the container, and
drawing one name out. This kind of manual process is used in lottery games. For
example in New Zealand we have Lotto in which 40 numbered balls are tossed around
in a container until one eventually drops out.

For a large population of people we could number everyone and then choose a
number at random using a computer. In this case we can obtain the probability of
getting a male using the law of probability relating to equally likely events, namely if
we have N equally likely outcomes of an experiment (the so-called sample space) and
M of these have a given characteristic, then the probability of choosing a member
with the given characteristic is simply M/N or p again. This means that for the

G. A. F. Seber, *Statistical Models for Proportions and Probabilities*,
SpringerBriefs in Statistics, DOI: 10.1007/978-3-642-39041-8_1,
© The Author(s) 2013

random experiment, p can now be regarded as a probability. If we then removed that person from the population and chose a second person at random from the remainder, then the probability of getting a male will be $(M-1)/(N-1)$ if the first choice was a male or $M/(N-1)$ if the first choice was a female. This is called sampling without replacement and the probability of choosing a male changes with each selection. However if N and M are large (say 10,000 and 4,800) then $p = 0.48$ and the respective probabilities for the second choice are 0.4799 and 0.48005, which could both be approximated by 0.48. If we carried on this sampling without replacement for say n times (the sample size), then provided n was a small enough fraction of N, the probability at each selection would remain approximately constant.

If we now changed the random experiment so that each selection is then returned to the population, then the experimental conditions would be the same for each selection so that the probability of choosing a male will be the same on each occasion, namely p. This sampling experiment is called sampling with replacement, and under certain conditions we saw above that sampling without replacement can be approximated by sampling with replacement. If we now change the problem to that of tossing a coin, then if the coin is perfectly balanced we would have two equally likely outcomes, a head and a tail. In this case the probability of getting a head is $\frac{1}{2}$ for each toss of the coin. However no coin is perfectly balanced so that the probability of getting a head will be an unknown probability p_H, say. How do we define this unknown? If we tossed the coin a very large number of times and tracked the proportion of heads we would see that this proportion would eventually settle down to a number which we can define as p_H.

We see from the above discussion that there is an interplay between the use of proportions and probabilities, and it is helpful to be clear as to which concept we are dealing with in constructing statistical models for experimental situations. We can therefore refer to p as a proportion or probability depending on the context. More generally, we may be interested in more than one proportion such as in responses to a questionnaire or voting preferences for a population. In what follows we focus on just a single proportion and consider some distribution theory and associated inference.

Suppose that the population under investigation is of size N as before, and M members are male, which we now call called the "marked" population so that $p = M/N$ is the proportion of marked in the population. If a random sample of size n is taken from the population without replacement (called a simple random sample or SRS)[1] and X is the number marked in the sample,[2] then X has a Hypergeometric distribution with probability function

$$f_1(x) = \Pr(X = x) = \binom{M}{x}\binom{N-M}{n-x} \bigg/ \binom{N}{n}, \qquad (1.1)$$

---

[1] The words "without replacement" are sometimes added to avoid ambiguity.

[2] We shall generally use capital letters (e.g., X and Y) for random variables, though this won't always be convenient, especially in sample survey theory.

where $\max\{n + M - N, 0\} \le x \le \min\{M, n\}$, though we usually have $x = 0, 1, \ldots, n$.[3] If $q = 1 - p$, it can be shown that an unbiased estimator of p is $\widehat{P} = X/n$ with variance

$$\sigma^2(\widehat{P}) = \text{var}(\widehat{P}) = \frac{pq}{n} r, \tag{1.2}$$

where

$$r = \frac{N - n}{N - 1} = 1 - \frac{n - 1}{N - 1} > 1 - f, \tag{1.3}$$

and $f = n/N$, the so-called sampling fraction. An unbiased estimate of the above variance is

$$\widehat{\sigma}^2(\widehat{P}) = \widehat{\text{var}}(\widehat{P}) = \frac{\widehat{P}(1 - \widehat{P})}{n - 1}(1 - f). \tag{1.4}$$

In inference, the focus is on the standard deviation σ and therefore on $\sqrt{1 - f}$, which takes the values $0.89, 0.95, 0.98$ when $f = 0.2, 0.1, 0.05$, respectively. This means that f can generally be neglected if $f < 0.1$ (or preferably $f < 0.05$) and we can then set $r = 1$ in (1.3).

If the sampling is with replacement, then, for each selection, the probability of selecting a marked individual is p and the proportion can now be treated as a probability. Since the selections or "trials" are mutually independent and n is fixed, X now has a Binomial distribution, denoted by $\text{Bin}(n, p)$ with probability function

$$f_2(x) = \binom{n}{x} p^x q^{n-x}, \quad x = 0, 1, \ldots, n.$$

We shall use the notation $X \sim \text{Bin}(n, p)$ and discuss this distribution in more detail in the next chapter. For this model we have

$$\text{E}(\widehat{P}) = p \quad \text{and} \quad \text{var}(\widehat{P}) = \frac{pq}{n}, \tag{1.5}$$

with unbiased variance estimate

$$\widehat{\text{var}}(\widehat{P}) = \frac{\widehat{P}(1 - \widehat{P})}{n - 1}. \tag{1.6}$$

We see then that if f can be ignored, we can approximate sampling without replacement by sampling with replacement, and approximate the Hypergeometric distribution by the Binomial distribution. This approximation can be established mathematically after some algebra using Stirling's inequality and the fact we have some large factorials, namely

$$\sqrt{\pi} n^{n+1/2} e^{-n+1/(12n+1)} < n! < \sqrt{\pi} n^{n+1/2} e^{-n+1/(12n)}.$$

---

[3] For further properties of the Hypergeometric distribution see Johnson et al. (2005: Chap. 6).

1.2 Inverse Sampling

Another method of estimating p is to use inverse sampling. Here sampling is continued without replacement until a fixed number of marked individuals, k, say, is obtained. The sample size n is now a the random variable, say Y, with probability function

$$f_3(y) = \frac{\binom{M}{k-1}\binom{N-M}{y-k}}{\binom{N}{y-1}} \cdot \frac{M-k+1}{N-y+1}$$

$$= \frac{\binom{y-1}{k-1}\binom{N-y}{M-k}}{\binom{N}{M}}, \quad y = k, k+1, \ldots, N+k-M,$$

which is the Negative (Inverse)-Hypergeometric distribution. This probability function can be expressed in a number of different ways (Johnson et al. 2005 , p. 255). We find that

$$\mathrm{E}(Y) = k\frac{N+1}{M+1} \quad \text{and} \quad \mathrm{var}(Y) = \frac{k(N+1)(N-M)(M+1-k)}{(M+1)^2(M+2)}.$$

Our usual estimate k/Y of $p = M/N$ is now biased. In estimating the inverse of a parameter for a discrete distribution we can often find a suitable estimate based on the inverse of a random variable with $+1$ or -1 added. We therefore consider

$$\widehat{P}_{in} = \frac{k-1}{Y-1}. \tag{1.7}$$

Now

$$\mathrm{E}(\widehat{P}_{in}) = \sum_{y=k}^{N+k-M} \frac{k-1}{y-1} \frac{\binom{y-1}{k-1}\binom{N-y}{M-k}}{\binom{N}{M}}$$

$$= \sum_{y-1=k-1}^{N+k-1-M} \frac{M}{N} \frac{\binom{y-2}{k-2}\binom{N-1-(y-1)}{M-1-(k-1)}}{\binom{N-1}{M-1}}$$

$$= p,$$

by setting $z = y - 1$ and using the result that $\sum_z f_3(z) = 1$ with new parameters $M-1, N-1$, and $k-1$. Hence $\widehat{P}_{in}$ is an unbiased estimate of p. In a similar fashion we can prove that

$$\mathrm{E}\left\{\frac{(k-1)(k-2)}{(Y-1)(Y-2)}\right\} = \frac{M(M-1)}{N(N-1)}. \tag{1.8}$$

To find a variance estimate it is helpful to define $\widehat{M} = N(k-1)/(Y-1)$ and consider

$$v_M = \frac{N^2(k-1)^2}{(Y-1)^2} - \frac{(k-1)(k-2)N(N-1)}{(Y-1)(Y-2)} - \frac{N(k-1)}{(Y-1)}. \tag{1.9}$$

Then, using (1.8),

$$E(v_M) = E(\widehat{M}^2) - M(M-1) - M$$
$$= E(\widehat{M}^2) - M^2 = \mathrm{var}(\widehat{M}).$$

(This method has been found useful for other discrete distributions, as we shall find below.) It can then be shown after some algebra that an unbiased estimate of $\mathrm{var}(\widehat{P}_{in})$ is

$$\widehat{\mathrm{var}}(\widehat{P}_{in}) = N^{-2}v_M$$
$$= \frac{(Y-k)(k-1)(N-Y+1)}{N(Y-1)^2(Y-2)}$$
$$= \left(1 - \frac{Y-1}{N}\right)\frac{\widehat{P}_{in}(1-\widehat{P}_{in})}{Y-2}. \tag{1.10}$$

Salehi and Seber (2001) derived the above unbiased estimates directly using a theorem due to Murthy (1957).

1.3 Application to Capture-Recapture

In the above discussion we have been interested in estimating $p = M/N$, where N is known. A common and effective method for estimating the unknown size N of an animal population is to mark or tag in some way M of the individuals, so that M is now known. An estimate of N is then $\widehat{N} = M/\widehat{P} = Mn/X$. This estimator is not only biased but the denominator can take a zero value. For simple random sampling, an approximately unbiased estimator is

$$N^* = \frac{(M+1)(n+1)}{(X+1)} - 1 \tag{1.11}$$

with an approximately unbiased variance estimator

$$v^* = \frac{(M+1)(n+1)(M-X)(n-X)}{(X+1)^2(X+2)}.$$

When $M+n \geq N$, both estimators are exactly unbiased (see Seber 1982, Sect. 3.1.1).

This method can also be applied in epidemiology to the problem of estimating the number N of people with a certain disease (e.g., diabetes) from two incomplete lists (Wittes and Sidel 1968). Each list "captures" a certain proportion of the N people with diabetes, but will miss some; we need to estimate those missed from both lists. In this case the first list (which "tags" the individuals by virtue of being on the first list) gives us $M (= n_1)$ people with diabetes while the second list give us $n (= n_2)$ people. Then, if the number on both lists is $m (= n_{11})$, our estimate of N from (1.10) now takes the form

$$N^* = \frac{(n_1 + 1)(n_2 + 1)}{(n_{11} + 1)} - 1.$$

This alternative notation is mentioned as it used in Sect. 5.1.3 for extending the problem to more than two lists. One of the key assumptions, namely that the two lists are independent, can be achieved if one of the lists is obtained through a random sample such as a random questionnaire (e.g., Seber et al. 2000).

In the case of inverse sampling for a capture-recapture experiment, with k being the predetermined number of marked and Y being the final random sample size, an unbiased estimator of N is

$$N_{in}^* = \frac{Y(M + 1)}{k} - 1$$

with exact variance (Seber 1982, Sect. 3.5)

$$\text{var}(N_{in}^*) = \frac{(M - k + 1)(N + 1)(N - M)}{k(M + 2)}.$$

1.4 Inference for a Proportion

In order to obtain a confidence interval for p we first construct a confidence interval for M and then divide it by N. In the case of simple random sampling without replacement, X is a discrete random variable so that it is not possible to construct an interval with an exact prescribed confidence so instead we focus on conservative confidence sets. Given $X = x$, this implies finding a set of values of y depending on x from the Hypergeometric probability function (1.1) that contains M with a confidence of at least $100(1 - \alpha)\%$. For example, a conservative two-sided confidence interval for M is (M_L, M_U), where M_L is the smallest integer M such that

$$\text{pr}(X \geq x) = \sum_{y \geq x} f_1(y) > \frac{\alpha}{2},$$

and M_U is the largest integer M such that

$$\text{pr}(X \leq x) = \sum_{y \leq x} f_1(y) > \frac{\alpha}{2}.$$

The above method of constructing a confidence interval is a standard one as it consists of the set of all M_0 such that a test of the hypothesis $H_0 : M = M_0$ versus the two-sided alternative $H_a : M \neq M_0$ is not rejected at the α level of significance. We essentially obtain the interval by "inverting" a family of hypothesis tests. Wendell and Schmee (2001) call this method the T-method, and the interval is usually shorter (cf. Buonaccorsi 1987) than the finite population version of the Clopper-Pearson interval found in Cochran (1977). They also introduced their L-method based on the likelihood function that appears to give intervals that are closer to the nominal confidence level than those from the T-method. The above theory can be readily applied to finding one-sided intervals. As iterative methods are available for computing various terms of the Hypergeometric distribution, methods based on using the exact values of the distribution rather than some approximation (e.g., by the Normal distribution mentioned below) are becoming readily available, and these are generally preferred. They can be used for any discrete distribution including distributions associated with inverse sampling. Methods using these exact values are referred to as "exact" methods, which is confusing. However this usage is common in the literature so I will continue to use it.

We saw above that the Hypergeometric distribution can be approximated by the Binomial distribution when the sampling fraction f is small enough. When p is small, another "rule of thumb" for the use of the approximation that is sometimes suggested is $p < 0.1$ and $n \geq 60$. The Binomial distribution and associated confidence intervals are discussed in detail in Chap. 2.

It is also of interest to see how we can use a Normal approximation to construct confidence intervals for p. A number of rules of thumb are available in the literature such as $N > 150$, $M > 50$, and $n > 50$, or $np > 4$, for which $\widehat{P}$ is approximately distributed as $N(p, \sigma^2(\widehat{P}))$. Using continuity corrections, an approximate two-sided $100(1 - \alpha)\%$ confidence interval is given by (p_L, p_U), where p_L and p_U satisfy

$$\mathrm{pr}(X \geq x \mid p = p_L) = 1 - \Phi \left(\frac{x - 0.5 - np_L}{\sqrt{np_L(1 - p_L)\frac{N-n}{N-1}}} \right) = \frac{\alpha}{2},$$

and

$$\mathrm{pr}(X \leq x \mid p = p_U) = \Phi \left(\frac{x + 0.5 - np_L}{\sqrt{np_U(1 - p_U)\frac{N-n}{N-1}}} \right) = \frac{\alpha}{2},$$

where $\Phi(x)$ is the distribution function for the standard Normal $N(0, 1)$ distribution. Let $z = z(\alpha/2)$ be the $100(1 - \frac{\alpha}{2})$ percentile of the standard Normal distribution. If $y = \frac{N-n}{N-1}z^2$, then setting $a = \pm z$ in $\Phi(a)$ for the above two equations, we have for example

$$ynp_L(1 - p_L) = (x - 0.5 - np_L)^2,$$

so that p_L is the smaller root of the quadratic

$$p^2(n^2 + ny) - p[ny + 2n(x - 0.5)] + (x - 0.05)^2.$$

Similarly p_U is the larger root of

$$p^2(n^2 + ny) - p[ny + 2n(x + 0.5)] + (x + 0.05)^2.$$

A crude approximation is to use the unbiased variance estimator $\widehat{\sigma}^2(\widehat{P})$ instead of $\sigma^2(\widehat{P})$ and, using a correction for continuity, we obtain the confidence interval

$$[p_L, p_U] = \left[\widehat{p} \pm z_{1-\frac{\alpha}{2}} \sqrt{\frac{N-n}{N(n-1)} \widehat{p}(1 - \widehat{p}) + \frac{1}{2n}} \right].$$

However this method is more of historical interest.

References

Buonaccorsi, J. P. (1987). A note on confidence intervals for proportions in finite populations. *The American Statistician, 41*, 215–218.

Cochran, W. G. (1977). *Sampling Techniques* (3rd edn.). New York: Wiley.

Murthy, M. N. (1957). Ordered and unordered estimators in sampling without replacement. *Sankhyā, 18*, 379–390.

Johnson, N.L., Kemp, A. W., & Kotz, S. (2005). *Univariate discrete distributions* (3rd edn.). New York: Wiley.

Salehi, M. M., & Seber, G. A. F. (2001). A new proof of Murthy's estimator which applies to sequential sampling. *Australian and New Zealand Journal of Statistics, 43*(3), 901–906.

Seber, G. A. F. (1982). Estimation of animal abundance and related parameters (2nd edn.). London: Griffin. Reprinted by Blackburn Press, Caldwell, New Jersey, U.S.A. (2002).

Seber, G. A. F., Huakau, J., & Simmons, D. (2000). Capture-recapture, epidemiology, and list mismatches: Two lists. *Biometrics, 56*, 1227–1232.

Wendell, J. P., & Schmee, J. (2001). Likelihood confidence intervals for proportions in finite populations. *The American Statistician, 55*, 55–61.

Wittes, J., & Sidel, V. W. (1968). A generalization of the simple capture-recapture model with applications to epidemiological research. *Journal of Chronic Diseases, 21*, 287–301.

Chapter 2
Single Probability

Abstract The Binomial distribution and its properties are discussed in detail including maximum likelihood estimation of the probability p. Exact and approximate hypothesis tests and confidence intervals are provided for p. Inverse sampling and the Negative Binomial Distribution are also considered.

Keywords Bernoulli trials · Maximum likelihood estimate · Likelihood-ratio test · Inverse sampling · Negative-Binomial distribution · Exact hypothesis test for a probability · Exact and approximate confidence intervals for a probability · Poisson approximation to the Binomial distribution

2.1 Binomial Distribution

An important model involving a probability is the Binomial distribution. It was mentioned in Chap. 1 as an approximation for the Hypergeometric distribution. It is, however, an important distribution in its own right as it arises when we have a fixed number n of Bernoulli experiments or "trials." Such trials satisfy the following three assumptions:

1. The trials are mutually independent.
2. Each trial has only two outcomes, which we can label "success" or "failure."
3. The probability p $(= 1 - q)$ of success is constant from trial to trial.

2.1.1 Estimation

If X is the number of successes from a fixed number n of Bernoulli trials, then X has the Binomial probability function

$$f_1(x) = \binom{n}{x} p^x q^{n-x}, \quad x = 0, 1, \ldots, n. \tag{2.1}$$

G. A. F. Seber, *Statistical Models for Proportions and Probabilities*,
SpringerBriefs in Statistics, DOI: 10.1007/978-3-642-39041-8_2,
© The Author(s) 2013

Using the Binomial Theorem we note that

$$\sum_{x=0}^{n} f_1(x) = (p+q)^n = 1.$$

Ignoring constants, the likelihood function is $L(p) = p^x q^{n-x}$ so that the maximum likelihood estimator of p is obtained by setting

$$\frac{\partial \log L(p)}{\partial p} = \frac{x}{p} - \frac{n-x}{1-p} = 0,$$

namely $\widehat{P} = x/n$. It is a maximum as the second derivative is negative. Also

$$-E\left[\frac{\partial^2 \log L(p)}{\partial p^2}\right] = E\left[\frac{X}{p^2} + \frac{n-X}{(1-p)^2}\right]$$
$$= \frac{n}{p} + \frac{n}{1-p}$$
$$= \frac{n}{pq},$$

which is the inverse of the Cramér-Rao lower bound. As $\widehat{P} = X/n$ is unbiased and has variance pq/n, it is the minimum variance unbiased estimator of p.

2.1.2 Likelihood-Ratio Test

To test the hypothesis $H_0 : p = p_0$ versus the alternative $H_a : p \neq p_0$ we can use the likelihood-ratio test

$$\Lambda_n = \frac{L(p_0)}{\sup_p L(\widehat{p})} = \frac{p_0^x (1-p_0)^{n-x}}{\widehat{p}^x (1-\widehat{p})^{n-x}}.$$

When H_0 is true, $-2\log \Lambda_n$ is asymptotically distributed as χ_1^2, the Chi-square distribution with one degree of freedom. We reject H_0 at the α level of significance if the observed value of $-2\log \Lambda_n$ exceeds $\chi_1^2(\alpha)$, the upper α tail value.

2.1.3 Some Properties of the Binomial Distribution

Moments of the Binomial distribution can be found by differentiating or expanding its moment generating function

$$M_x(t) = E(e^{tx}) = (q + e^t p)^n.$$

For example, $E(X) = M'_x(0) = np$ and $E(X^2) = \frac{1}{2!}M''_X(0) = n(n-1)p^2 + np$. Factorial moments are sometimes useful such as (setting $y = x - r$)

$$E[X(X-1)\cdots(X-r+1)] = \sum_{x=r}^{n} x(x-1)\cdots(x-r+1)\frac{n!}{x!(n-x)!}p^x q^{n-x}$$

$$= p^r n \cdots (n-r+1) \sum_{y=0}^{n-r} \binom{n-r}{y} p^y q^{n-r-y}$$

$$= p^r n(n-1)\cdots(n-r+1)(p+q)^{n-r}$$

$$= p^r n(n-1)\cdots(n-r+1).$$

For example, setting $r = 2$, $E[X(X-1)] = p^2 n(n-1)$.

One other result that has been found useful is in the situation of studying variables like $\widehat{P}^{-1} = n/X$ as an estimate of p^{-1}. This raises problems as we can have $X = 0$. A useful idea is to modify the variable and consider

$$E\left(\frac{n+1}{X+1}\right) = \frac{1}{p} \sum_{i=0}^{n} \frac{n+1!}{(x+1)!(n+1-x-1)!} p^{x+1} q^{n+1-(x+1)}$$

$$= \frac{1}{p} \sum_{y=1}^{n+1} \binom{n+1}{y} p^y q^{n+1-y} \quad (y = x+1)$$

$$= \frac{1}{p}[(p+q)^{n+1} - q^{n+1}]$$

$$= \frac{1}{p}(1 - q^{n+1}).$$

For large n, $(n+1)/(X+1)$ is an approximately unbiased estimate of $1/p$. This technique works well with a number of other discrete distributions as we saw in Chap. 1.

When $n = 1$, X becomes an indicator variable J, say, where $J = 1$ with probability p, and $J = 0$ with probability q. Then $E(J) = p$ and

$$\mathrm{var}(J) = E(J^2) - (E(J))^2 = p^2 - p = pq.$$

If J_i is the indicator variable associated with the ith trial, we can now write $X = \sum_{i=1}^{n} J_i$ with mean np and variance $\sum_{i=1}^{n} \mathrm{var}(J_i) = npq$. Also, $\widehat{P} = X/n = \bar{J}$ so that by the Central Limit Theorem $\widehat{P}$ is asymptotically $N(p, pq/n)$.

2.1.4 Poisson Approximation

If we let $p \to 0$ and $n \to \infty$ such that $\lambda_n = np \to \lambda$, where λ is a constant, then the Binomial moment generating function is given by

$$(q + pe^t)^n = \left[1 - \frac{\lambda_n}{n}(1 - e^t)\right]^n \to e^{-\lambda(1 - e^t)},$$

the moment generating function of the Poisson distribution, Poisson(λ), with mean λ. We see then that the Binomial distribution can be approximated by the Poisson distribution with mean np when p is small and n is large. Quantiles of the Poisson distribution can be obtained from the Chi-square distribution using the result

$$\mathrm{pr}(Y \le x) = \mathrm{pr}(\chi^2_{2(1+x)} \le 2\lambda),$$

where $Y \sim$ Poisson(λ). We could use this result to construct an approximate confidence interval for np and hence for p.

2.2 Inverse Sampling

Suppose we have a sequence of Bernoulli trials that continues until we have r successes. If W is the number of failures, then the sample size $Y = W + r$ is random with the last trial being a success. Hence, W has a Negative-Binomial distribution with probability function

$$f_2(w) = \binom{w + r - 1}{r - 1} p^{r-1} q^w \cdot p = \binom{w + r - 1}{r - 1} p^r q^w, \quad w = 0, 1, \ldots.$$

This probability function can be expressed in a number of different ways (Johnson et al. 2005, Chap. 5). The moment generating function of W is $M_w(t) = (Q_0 - P_0 e^t)^{-r}$, where

$$P_0 = \frac{1 - p}{p} \quad \text{and} \quad Q_0 = \frac{1}{p}.$$

Differentiating $M_w(t)$ leads to

$$\mathrm{E}(W) = rP_0 \quad \text{and} \quad \mathrm{var}(W) = rP_0 Q_0.$$

We now wish to find an unbiased estimator of p and an unbiased estimate of its variance. We do this using estimators due to Murthy (1957) that were shown by Salehi and Seber (2001) to apply to inverse sampling. Since W is a complete sufficient statistic for p, it can be shown that the minimum variance unbiased estimator for p is

$$\widehat{P}_{in} = \frac{r-1}{r+W-1} = \frac{r-1}{Y-1},$$

which is the same as for sampling without replacement (Sect. 1.2). This is perhaps not surprising as the same equality occurs with simple random sampling with or without replacement. An unbiased variance estimator of $\mathrm{var}(\widehat{P}_{in})$ is (Salehi and Seber 2001)

$$\widehat{\mathrm{var}}(\widehat{P}_{in}) = \frac{\widehat{P}_{in}(1-\widehat{P}_{in})}{Y-2}.$$

Unbiasedness can also be proved directly using the methods of Sect. 1.2.

2.3 Inference for a Probability

There is a considerable literature on confidence intervals for the Binomial distribution. We begin by considering so-called "exact" confidence intervals mentioned in Sect. 1.4, which are confidence intervals based on the exact Binomial distribution and not on an approximation for it. This also leads to an exact hypothesis test. Because of the discreteness of the distribution we cannot normally obtain a confidence interval with an exact prescribed confidence of $(1-\alpha)\,\%$ but rather we aim for a (conservative) confidence level of at least $100(1-\alpha)\,\%$. After considering exact intervals we will then derive some approximate intervals and tests based on approximations for the Binomial distribution.

2.3.1 Exact Intervals

Given an observed value $X = x$ for a Binomial distribution, we can follow the method described in Sect. 1.4 to obtain an exact confidence interval. We want to find probabilities p_L and p_U such that, for a two-sided confidence interval with confidence $100(1-\alpha)\,\%$,

$$\mathrm{pr}(X \geq x \mid p = p_L) = \sum_{i=x}^{n} \binom{n}{i} p_L^i (1-p_L)^{n-i} = \frac{\alpha}{2},$$

and

$$\mathrm{pr}(X \leq x \mid p = p_U) = \sum_{i=0}^{x} \binom{n}{i} p_U^i (1-p_U)^{n-i} = \frac{\alpha}{2}.$$

The interval (p_L, p_U) is known as the Clopper-Pearson confidence interval (Clopper and Pearson 1934). The tail of the Binomial distribution can be related to the tail of the F-distribution through the relationship (Jowett 1963)

$$\sum_{i=0}^{x} \binom{n}{i} p^i (1-p)^{n-i} = \text{pr}\left\{ Y \le \frac{(1-p)(x+1)}{p(n-x)} \right\}, \qquad (2.2)$$

where Y has the F-distribution $F(2(n-x), 2(x+1))$. If $F_{1-\frac{\alpha}{2}}(\cdot, \cdot)$ denotes the $100(1-\alpha/2)$th percentile of the F-distribution, then[1]

$$p_L = \frac{x}{x + (n-x+1)F_{1-\frac{\alpha}{2}}(2(n-x+1), 2x)}$$

and

$$p_U = \frac{(x+1)F_{1-\frac{\alpha}{2}}(2(x+1), 2(n-x))}{n-x + (x+1)F_{1-\frac{\alpha}{2}}(2(x+1), 2(n-x))}.$$

The percentiles of the F-distribution are provided by a number of statistical software packages. The above interval is very conservative and the coverage probability often substantially exceeds $1-\alpha$. One way of dealing with this is to use the so-called "mid p-value" where only half of the probability of the observed result is added to the probability of more extreme results (Agresti 2007, pp. 15–16). This method is particularly useful for very discrete distributions (i.e., with few well-spaced observed values). One theoretical justification for its use is given by Hwang and Yang (2001). Further comments about the method are made by Berry and Armitage (1995).

2.3.2 Exact Hypothesis Test

There is some controversy as to how to carry out an "exact" test of the hypothesis $H_0 : p = p_0$ versus the two-sided alternative $H_a : p \ne p_0$. One method is as follows. If $\widehat{p} < p_0$, we evaluate

$$g(p_0) = \sum_{i=0}^{x} \binom{n}{i} p_0^i (1-p_0)^{n-i} = \gamma,$$

where 2γ is the p-value of the test. If $\widehat{p} > p_0$, we evaluate $1-g(p_0)+\text{pr}(X=x) = \delta$, where 2δ is the p-value of the test. We can use (2.2) to evaluate the p-value exactly or use F-tables if the software is not available. This method is referred to as the TST or Twice the Smaller Tail method by Hirji (2006, p. 59). As p-values tend to be too large, some statisticians prefer to use the mid p-value, as mentioned above. It involves halving the probability of getting the observed value x under the assumption of H_0 being true. Hirji (2006, pp. 70–73) also defines three other methods for carrying out the test,[2] and discusses exact tests in general.

---

[1] See, for example, http://www.ppsw.rug.nl/~boomsma/confbin.pdf.

[2] See also Fay (2010).

2.3.3 Approximate Confidence Intervals

We saw above that $\widehat{P}$ is asymptotically $N(p, pq/n)$, or more rigorously,

$$\frac{\widehat{P} - p}{\sqrt{pq/n}} \text{ is asymptotically } N(0, 1).$$

An approximate two-sided $100(1 - \alpha)\%$ confidence interval for p therefore has upper and lower limits that are solutions of the quadratic equation

$$(\widehat{p} - p)^2 = z(\alpha/2)^2 \frac{p(1 - p)}{n}, \tag{2.3}$$

where $z(\alpha/2)$ is the $\alpha/2$ tail value of standard Normal N(0,1) distribution. This confidence interval is usually referred to as the score confidence interval, though it is also called the Wilson interval introduced in 1927 as it inverts the test $H_0 : p = p_0$ obtained by substituting p for p_0 in (2.3). We note for later reference that $z(\alpha/2)^2 = \chi_1^2(\alpha)$, where the latter is the $1 - \alpha$ quantile of the Chi-squared distribution with one degree of freedom.

An alternative method is based on the fact that

$$\frac{\widehat{P} - p}{\sqrt{\widehat{P}(1 - \widehat{P})/n}} \text{ is also asymptotically } N(0, 1)$$

yielding the confidence interval

$$\widehat{p} \pm z(\alpha/2)\sqrt{\frac{\widehat{p}(1 - \widehat{p})}{n}}. \tag{2.4}$$

This is usually referred as the Wald confidence interval for p, since it results from inverting the Wald test for p. It is the set of p_0 values having a p-value exceeding α in testing $H_0 : p = p_0$ versus $H_a : p \neq p_0$ using the test statistic $z = (\widehat{p} - p_0)/\sqrt{\widehat{p}(1 - \widehat{p})/n}$.

Agresti and Coull (1998) compared the above two methods and the exact confidence interval and came to a number of conclusions. First, the score interval performed the best in having coverage probabilities close to the nominal confidence level. Second, they recommended its use with nearly all sample sizes and parameter values. Third, the exact interval remains quite conservative even for moderately large sample sizes when p tends to be 0 or 1. Fourth, the Wald interval fairs badly when p is near 0 or 1, one reason being that $\widehat{p}$ is used as the midpoint of the interval when the Binomial distribution is highly skewed. Finally, they provided an adaption of the Wald interval with $\widehat{p} = x/n$ replaced by $(x + 2)/(n + 4)$ in (2.4) (the "add two successes and two failures" rule) that also performs well even for small samples when $\alpha = 0.05$.

Brown et al. (2001) discussed the extreme volatility and oscillation of the Wald interval's behavior, due to the discreteness of the Binomial distribution, even when n is quite large and p is not near 0 or 1. They showed that the usual rules given in texts for when the Wald interval is satisfactory (e.g., $np, n(1 - p)$ are ≥ 5 (or 10)) are somewhat defective and recommended three intervals: (1) the score (Wilson) interval, (2) an adjusted Wald interval they call the Agresti-Coull interval, and (3) an interval based on Jeffrey's prior distribution for p that they call Jeffrey's interval.

Solving the quadratic (2.3), the Wilson interval can be put in the form, after some algebra,

$$p \in \tilde{p} \pm \frac{\kappa n^{1/2}}{n + \kappa^2} (\widehat{pq} + \kappa^2/(4n))^{1/2},$$

where

$$\kappa = z(\alpha/2), \quad \text{and} \quad \tilde{p} = \frac{x + \kappa^2/2}{n + \kappa^2} = \frac{\tilde{x}}{\tilde{n}}, \text{ say.}$$

The Agresti-Coull interval takes the form

$$p \in \tilde{p} \pm \kappa \sqrt{\frac{\tilde{p}(1 - \tilde{p})}{\tilde{n}}}.$$

Note that $z(0.025) = 1.96 \approx 2$, which gives the above "add 2" rule. These two intervals have the same recentering that can increase coverage significantly for p away from 0 or 1 and eliminate systematic bias. Further simulation support for the "add 2" rule is given by Agresti and Caffo (2000).

Using a Beta prior distribution for p, say Beta(a_1, a_2) (a conjugate prior for the Binomial distribution), it can be shown that the posterior distribution for p is Beta$(a_1 + x, a_2 + n - x)$. Using Jeffrey's prior $(a_1 = 1/2, a_2 = 1/2)$, a $100(1 - \alpha)\%$ equal-tailed posterior confidence interval $(p_L(x), p_U(x))$ is given by

$$p_L(x) = B\left(\frac{\alpha}{2}; x + \frac{1}{2}, n - x + \frac{1}{2}\right) \text{ and } p_U(x) = B\left(1 - \frac{\alpha}{2}; x + \frac{1}{2}, n - x + \frac{1}{2}\right).$$

where $p_L(0) = 0$, $p_U(n) = 1$, and $B(\gamma; m_1, m_2)$ is the γ quantile of the Beta(m_1, m_2) distribution. Brown et al. (2001) considered further modifications when p is near 0 or 1. They recommended either the Wilson or Jeffrey's intervals for $n \leq 40$, while all three are fairly similar for $n > 40$, with the Agresti-Coull interval being easy to use and remember, though a little wider.

Brown et al. (2002) added as a contender the interval obtained by inverting the likelihood-ratio test that accepts the null hypothesis $H_0 : p = p_0$ of Sect. 2.1.2 at the α level of significance, though it requires some computation. It takes the form

$$\left\{ p : x\log p + (n - x)\log (1 - p) \geq x\log \widehat{p} + (n - x)\log (1 - p) - \chi_1^2\left(\frac{\alpha}{2}\right) \right\}.$$

They also concluded that the Wilson, the likelihood ratio, and Jeffrey's intervals are comparable in both coverage and length, though the Jeffrey's interval is a bit shorter on average. Further comments are made by Newcombe (1998), who compared a number of intervals.

References

Agresti, A. (2007). *An introduction to categorical data analysis*. Hoboken, NJ: Wiley Interscience.

Agresti, A., & Caffo, B. (2000). Simple and effective confidence intervals for proportions and differences of proportions result from adding two successes and two failures. *The American Statistician, 54*(4), 280–288.

Agresti, A., & Coull, B. A. (1998). Approximate is better than 'exact' for interval estimation of binomial proportions. *The American Statistician, 52*(2), 119–126.

Berry, G., & Armitage, P. (1995). Mid-P confidence intervals: A brief review. *The Statistician, 44*(4), 417–423.

Brown, L. D., Cai, T. T., & DasGupta, A. (2001). Interval estimation of a binomial proportion. *Statistical Science 16*(2), 101–133.

Brown, L. D., Cai, T. T., & DasGupta, A. (2002). Confidence intervals for a binomial proportion and asymptotic expansions. *The Annals of Statistics, 30*(1), 160–210.

Clopper, C., & Pearson, E. S. (1934). Ordered and unordered estimators in sampling without replacement. *Sankhyā, 18*, 379–390.

Fay, M. P. (2010). Two-sided exact tests and matching confidence intervals for discrete data. *The R Journal 2*, 53–58. (See also http://journal.rproject.org/archive/2010-1/RJournal_2010-1_Fay.pdf)

Hirji, K. F. (2006). *Exact analysis of discrete data*. Boca Raton, FL: Taylor and Francis.

Hwang, J. T. G., & Yang, M.-C. (2001). An optimality theory for mid p-values in 2×2 contingency tables. *Statistica Sinica, 11*, 807–826.

Johnson, N. L., Kemp, A. W., & Kotz, S. (2005). *Univariate discrete distributions* (3rd ed.). New York: Wiley.

Jowett, D. H. (1963). The relationship between the binomial and F distributions. *The Statistician, 3*(1), 55–57.

Murthy, M. N. (1957). Ordered and unordered estimators in sampling without replacement. *Sankhyā, 18*, 379–390.

Newcombe, R. G. (1998). Two-sided confidence intervals for the single proportion: Comparison of seven methods. *Statistics in Medicine, 17*(8), 857–872.

Salehi, M. M., & Seber, G. A. F. (2001). A new proof of Murthy's estimator which applies to sequential sampling. *Australian and New Zealand Journal of Statistics, 43*(3), 901–906.

Chapter 3
Several Proportions or Probabilities

Abstract We discuss the Multi-hypergeometric and Multinomial distributions and their properties with the focus on exact and large sample inference for comparing two proportions or probabilities from the same or different populations. Relative risks and odds ratios are also considered. Maximum likelihood estimation, asymptotic normality theory, and simultaneous confidence intervals are given for the Multinomial distribution. The chapter closes with some applications to animal populations, including multiple-recapture methods, and the delta method.

Keywords Multi-hypergeometric distribution · Multinomial Distribution · Comparing two proportions or probabilities · Relative risk · Odds ratio · Maximum likelihood estimation · Simultaneous confidence intervals for probabilities · Random distribution of animals · Multiple-recapture models · Delta method

3.1 Multi-Hypergeometric Distribution

Suppose we have k subpopulations of sizes M_i ($i = 1, 2, \ldots, k$), where $\sum_{i=1}^{k} M_i = N$, the total population size. Let $p_i = M_i/N$. A simple random sample of size n is taken from the population yielding X_i from the ith subpopulation ($i = 1, 2, \ldots, k$). The joint probability function of $\mathbf{X} = (X_1, X_2, \ldots, X_k)'$ is the Multi-hypergeometric distribution (see, for example, Johnson et al. 1997, Chap. 39), namely

$$f(\mathbf{x}) = \mathrm{pr}(\mathbf{X} = \mathbf{x}) = \prod_{i=1}^{k} \binom{M_i}{x_i} \bigg/ \binom{N}{n}, \quad 0 \le x_i \le \min(n, M_i), \sum_{i=1}^{k} x_i = n.$$

(3.1)

Since we can add the subpopulations together we see that the marginal distribution of an X_i is also Hypergeometric, with two subpopulations M_i and $N - M_i$, namely

G. A. F. Seber, *Statistical Models for Proportions and Probabilities*,
SpringerBriefs in Statistics, DOI: 10.1007/978-3-642-39041-8_3,
© The Author(s) 2013

$$f_i(x_i) = \binom{M_i}{x_i}\binom{N - M_i}{n - x_i}\bigg/\binom{N}{n}.$$

In a similar fashion we see that the probability function of $X_1 + X_2$ is the Multi-hypergeometric distribution

$$f_{12}(x_1, x_2) = \binom{M_1 + M_2}{x_1 + x_2}\binom{N - M_1 - M_2}{n - x_1 - x_2}\bigg/\binom{N}{n}.$$

From Sect. 1.1, $\mathrm{var}(X_i) = np_i(1 - p_i)r$, where $r = (N - n)/(N - 1)$, and

$$\mathrm{var}(X_1 + X_2) = nr(p_1 + p_2)(1 - p_1 - p_2).$$

To find the covariance, $\mathrm{cov}(X_1, X_2)$, of X_1 and X_2 directly requires some awkward algebra. Fortunately we can use the following result:

$$\begin{aligned}
\mathrm{cov}(X_1, X_2) &= \frac{1}{2}\{\mathrm{var}(X_1 + X_2) - \mathrm{var}(X_1) - \mathrm{var}(X_2)\} \\
&= rn\{(p_1 + p_2)(1 - p_1 - p_2) - p_1(1 - p_2) - p_2(1 - p_2)\} \\
&= -rnp_1p_2. \tag{3.2}
\end{aligned}$$

We then find that if $q_i = 1 - p_i$, then

$$\begin{aligned}
\mathrm{var}(X_1 - X_2) &= \mathrm{var}(X_1) + \mathrm{var}(X_2) - 2\mathrm{cov}(X_1, X_2) \\
&= nr\{p_1q_1 + p_2q_2 + 2p_1p_2\} \\
&= nr\{p_1 + p_2 - (p_1 - p_2)^2\}. \tag{3.3}
\end{aligned}$$

3.2 Comparing Two Proportions from the Same Population

We consider two different situations depending on whether the proportions are based on independent data or correlated data. The latter situation involving dependent data is a special case of matched-pairs discussed further in Sect. 4.4.4.

3.2.1 Nonoverlapping Proportions

Suppose we wish to estimate the difference in the proportions $p_1 - p_2$ voting for two candidates in an election with N voters. Let Y_i be a random variable that takes the value $+1$ if the sampled person prefers candidate one, -1 for candidate two, and 0 otherwise. Then, from Scott and Seber (1983), an estimator of $p_1 - p_2$ is

$$\widehat{P}_1 - \widehat{P}_2 = \frac{X_1 - X_2}{n} = \overline{Y},$$

so that by the Central Limit Theorem,

$\widehat{P}_1 - \widehat{P}_2$ is asymptotically $N(p_1 - p_2, r\{p_1 + p_2 - (p_1 - p_2)^2\}/n)$,

where $r = (N-n)/(N-1)$. We can use the above result to make large sample inferences about $p_1 - p_2$. For example, replacing the p_i by their estimates, an approximate 95 % confidence interval for $p_1 - p_2$ is

$$\widehat{p}_1 - \widehat{p}_2 \pm 1.96\{r[\widehat{p}_1 + \widehat{p}_2 - (\widehat{p}_1 - \widehat{p}_2)^2]/n\}^{1/2}. \tag{3.4}$$

To test $H_0 : p_1 = p_2$ we can use the Normal approximation

$$z_0 = \frac{\widehat{p}_1 - \widehat{p}_2}{\sqrt{r(\widehat{p}_1 + \widehat{p}_2)/n}}, \tag{3.5}$$

where z_0 is from $N(0, 1)$. For a two-sided alternative $H_a : p_1 \neq p_2$ we reject H_0 at the α level of significance if $|z_0| > z(\alpha/2)$, where α is the upper α point of the $N(0, 1)$ distribution.

3.2.2 Dependent Proportions

There is another common situation where the two proportions are dependent and overlap (Wild and Seber 1993). Suppose we have a population of N people and a sample of size n is chosen at random without replacement. Each selected person is asked two questions to each of which they answer "Yes" (1) or "No" (2), so that p_{12} is the proportion answering yes to the first question and no to the second, p_{11} is the proportion answering yes to both questions, and so forth. Then the proportion answering yes to the first question is $p_1 = p_{11} + p_{12}$ and the proportion answering yes to the second question is $p_2 = p_{11} + p_{21}$. Let X_{ij} ($i = 1, 2; j = 1, 2$) be the number observed in the sample in the category with probability p_{ij}, let $X_1 = X_{11} + X_{12}$, the number answering yes to the first question, and let $X_2 = X_{11} + X_{21}$ be the number answering yes to the second question. The interest is then in comparing p_1 and p_2, and in the popular press the fact that they overlap on p_{11} is often ignored in the calculations. For example, only $\widehat{p}_1 = x_1/n$ and $\widehat{p}_2 = x_2/n$ are reported while x_{12} and x_{21} are not.

The four variables X_{ij} have a Multi-hypergeometric distribution, and

$$\widehat{P}_1 - \widehat{P}_2 = \frac{X_1 - X_2}{n} = \frac{X_{12} - X_{21}}{n} = \widehat{P}_{12} - \widehat{P}_{21}.$$

Then

$$E(\widehat{P}_1 - \widehat{P}_2) = p_{12} - p_{21} = p_1 - p_2,$$

and, from (3.3),

$$\begin{aligned}
\mathrm{var}(\widehat{P}_1 - \widehat{P}_2) &= \frac{1}{n^2}\mathrm{var}(X_{12} - X_{21}) \\
&= \frac{1}{n^2}\{\mathrm{var}(X_{12}) + \mathrm{var}(X_{21}) - 2\mathrm{cov}(X_{12}, X_{21})\} \\
&= \frac{r}{n}\{p_{12} + p_{21} - (p_{12} - p_{21})^2\}.
\end{aligned} \tag{3.6}$$

We can now use the above theory to make inferences about $p_1 - p_2$. However, if we are only given the values $\widehat{p}_1$ and $\widehat{p}_2$, which is often the case with news reports, what can we do? Using $p_{11} + p_{12} + p_{21} + p_{22} = 1$, we first note that

$$p_{12} + p_{21} = p_1 + p_2 - 2p_{11} = q_1 + q_2 - 2p_{22}, \quad q_i = 1 - p_i \quad (i = 1, 2).$$

If $d = |p_{12} - p_{21}|$, it follows from (3.6) and $p_{12} + p_{21} > |p_{12} - p_{21}|$ that

$$\frac{r}{n}d(1 - d) \le \mathrm{var}(\widehat{P}_1 - \widehat{P}_2) \le \frac{r}{n}[\min\{(p_1 + p_2), (q_1 + q_2)\} - (p_1 - p_2)^2].$$

Estimating p_i by $\widehat{P}_i$ and d by $|\widehat{P}_1 - \widehat{P}_2|$ we can use the estimated upper bound to construct an approximate conservative confidence interval for $p_1 - p_2$, as in (3.4).

There may be a problem with the unknown responses "Don't know" and "Non-responses." If we lump the unknown responses with the "No" responses, then we are only looking at the proportions of people who are prepared to make a "Yes" response. We could also include the unknown responses with the "Yes" responses in one question and with the "No" responses in the other so as to minimize the difference $\widehat{P}_1 - \widehat{P}_2$. If this difference is still significant, then we have stronger evidence of a real difference.

If we wish to test $H_0 : p_1 = p_2$ we set $p_{12} = p_{21}$ and use (3.5), namely

$$z_0 = \frac{\widehat{p}_{12} - \widehat{p}_{21}}{\sqrt{r(\widehat{p}_{12} + \widehat{p}_{21})/n}} \tag{3.7}$$

A lower bound for z_0 is

$$|z_L| = \frac{|\widehat{p}_1 - \widehat{p}_2|}{\sqrt{r \cdot \min(\widehat{p}_1 + \widehat{p}_2, \widehat{q}_1 + \widehat{q}_2)/n}}.$$

3.2.3 Two Independent Proportions

Suppose we have simple random samples of size n_i from two independent Hypergeometric distributions with unknown proportions $p_i = M_i/N_i$, sampling fractions $f_i = n_i/N_i$, and estimators $\widehat{P}_i = X_i/n_i$. If $r_i = (N_i - n_i)/(N_i - 1)$, then, by the Central Limit Theorem,

$$\frac{\widehat{P}_1 - \widehat{P}_2 - (p_1 - p_2)}{\sqrt{r_1 \widehat{P}_1(1 - \widehat{P}_1)/n_1 + r_2 \widehat{P}_2(1 - \widehat{P}_2)/n_2}}$$

is approximately Normally distributed as $N(0, 1)$ when $n_i p_i$, $n_i q_i$, and N are sufficiently large. There are problems with this approximation when the f_i are large, say greater than 0.5 (Lahiri et al. 2007). In most applications, $f_i < 0.1$ so that Binomial approximations can be used along with the theory in the next section.

To test $H_0 : p_1 = p_2 (= p$, say $)$ we can use the above Normal approximation with the common value of p now estimated by $(X_1 + X_2)/(n_1 + n_2)$, giving a score test. This test was compared with a bootstrap type of test called the E-test by Krishnamoorthy and Thomson (2002) that involves computing sums of Hypergeometric probabilities.

3.3 Comparing Two Probabilities from Independent Binomial Distributions

We shall look at several different comparisons such as the difference, ratio (odds) and odds ratio (defined below) for two probabilities. Estimation, hypothesis testing, and confidence intervals, including exact procedures will be considered.

3.3.1 Difference of Two Probabilities

Like the single probability, this topic has received a great deal of attention. Suppose we have independent samples from independent Binomial distributions $\text{Bin}(n_i, p_i)$ $(i = 1, 2)$ and we wish to make inferences about $\delta = p_1 - p_2$. If the number of "successes" are respectively X_1 and X_2, then we have the estimator

$$\widehat{\delta} = \widehat{P}_1 - \widehat{P}_2 = \frac{X_1}{n_1} - \frac{X_2}{n_2}$$

with variance

$$\text{var}(\widehat{\delta}) = \text{var}(\widehat{P}_1) + \text{var}(\widehat{P}_2) = \frac{p_1 q_1}{n_1} + \frac{p_2 q_2}{n_2}.$$

Since $\widehat{P}_i$ is asymptotically Normal, the difference is also asymptotically Normal for large n_i. Replacing the p_i by their estimates in $\text{var}(\widehat{\delta})$ we have the usual $100(1-\alpha)\%$ confidence interval

$$p_1 - p_2 \in (\widehat{p}_1 - \widehat{p}_2) \pm \kappa \sqrt{\frac{\widehat{p}_1(1-\widehat{p}_1)}{n_1} + \frac{\widehat{p}_2(1-\widehat{p}_2)}{n_2}}, \qquad (3.8)$$

where $\kappa = z(\alpha/2)$. This interval can be obtained by inverting the so-called Wald test for the hypothesis $H_0 : p_1 = p_2 (= p,$ say$)$ and is sometimes referred to as the Wald interval. This interval has poor coverage properties and a number of alternatives have been suggested that we now consider.

Agresti and Caffo (2000) suggested replacing each $\widehat{p}_i$ by p_i^* and n_i by n_i^* in (3.8), where

$$p_i^* = \frac{x_i^*}{n_i^*} = \frac{x_i + \kappa^2/2}{n_i + \kappa^2} \approx \frac{x_i + 2}{n_i + 4},$$

for 95% confidence intervals, their so-called "add 2" rule. This performs well over a wide range of parameter values and is easy to remember and compute. Because of the variety of recommended (and uncertain) sample size guidelines for the Wald intervals, Agresti and Caffo suggest that with their method, "one might simply by-pass sample size rules." Agresti and Min (2005, p. 519) found that the above method "performs well in a broad variety of conditions, even with small samples." They also found that it performs at least as well as the Bayesian intervals, and usually better, in terms of "the prevalence of under-coverage probabilities" when the samples are small. The interval does tend to be wider though. A number of authors, in their comparisons, refer to another version of the Agresti and Caffo interval that uses an "add 1" rule.

Another method is the hybrid-score interval of Newcombe (1998, method 10), which is computationally straightforward and performs well (cf. Brown and Li 2005). From Eq. (2.3) with for p replaced by p_1 and p_2, respectively, we let (ℓ_i, u_i) be the lower and upper roots of

$$(\widehat{p}_i - p_i)^2 = z(\alpha/2)^2 \frac{p_i(1-p_i)}{n}.$$

Then the lower and upper bounds (p_L, p_U) of Newcombe's $100(1-\alpha)\%$ confidence interval are

$$p_L = (\widehat{p}_1 - \widehat{p}_2) - z(\alpha/2) \sqrt{\frac{\ell_1(1-\ell_2)}{n_1} + \frac{u_2(1-u_2)}{n_2}}$$

and

$$p_U = (\widehat{p}_1 - \widehat{p}_2) + z(\alpha/2) \sqrt{\frac{u_1(1-u_2)}{n_1} + \frac{\ell_2(1-\ell_2)}{n_2}}.$$

Roughly speaking, the method uses average boundary variance estimates.

There is one more method that is more computationally complex and requiring an iterative method, which can be found by inverting the score test of the hypothesis $H_0 : \delta = d$. Although this method performs no better than the previous closed form methods, it is instructive to see how the inversion process works. Following Mee (1984), we first find the maximum likelihood estimates $\widetilde{p}_i$ of the p_i subject to the constraint $p_1 - p_2 = d$. The log likelihood function is the sum of the two log likelihoods to which we add $\lambda(p_1 - p_2 - d)$, where λ is a Lagrange multiplier. Differentiating this expression with respect to p_1 and p_2 we get the maximum likelihood equations

$$\frac{x_1}{\widetilde{p}_1} - \frac{n_1 - x_1}{1 - \widetilde{p}_1} + \lambda = 0, \quad \frac{x_2}{\widetilde{p}_2} - \frac{n_2 - x_2}{1 - \widetilde{p}_2} - \lambda = 0, \quad \text{and} \quad \widetilde{p}_1 - \widetilde{p}_2 = d.$$

Setting $\widehat{p}_i = x_i/n_i$, we find that these equations reduce to

$$\frac{\widetilde{p}_1 - \widehat{p}_1}{\widetilde{v}_1} + \frac{\widetilde{p}_2 - \widehat{p}_2}{\widetilde{v}_2} = 0 \quad \text{and} \quad \widetilde{p}_2 = \widetilde{p}_1 - d,$$

where $\widetilde{v}_i = \widetilde{p}_i(1 - \widetilde{p}_i)/n_i$. The confidence interval is then given by

$$\{d : |\widehat{p}_1 - \widehat{p}_2|/(\widetilde{v}_1 + \widetilde{v}_2)^{1/2} < z(\alpha/2)\}.$$

Mee presents the $\widetilde{p}_i$ as unique solutions of the following cubic equations (for $0 < \widetilde{p}_i < 1$, $i = 1, 2$, and d replaced by $-d$ as he works with $p_2 - p_1$):

$$\widetilde{p}_1 = \frac{\widehat{p}_1 \widetilde{v}_1^{-1} + (\widehat{p}_2 + d)\widetilde{v}_2^{-1}}{\widetilde{v}_1^{-1} + \widetilde{v}_2^{-1}} \quad \text{and} \quad \widetilde{p}_2 = \widetilde{p}_1 - d,$$

subject to the constraints $0 < \widetilde{p}_i < 1$ ($i = 1, 2$). Iterative methods are proposed to find the endpoints of the confidence interval and Mee suggests using the endpoints of the Wald confidence interval as starting points. Closed form expressions are given by Miettinen and Nurminen (1985) and discussed by Andrés and Tejedor (2002).

One parameter that has raised some interest is medicine is "the number needed to treat" or $NNT = 1/(p_1 - p_2)$. A confidence interval for it can be found, for example, by inverting an interval for $p_1 - p_2$, provided the latter does not contain zero (cf. Fagerland et al. 2011).

We can test the hypothesis $H_0 : p_1 = p_2 (= p$, say) using the large sample score test

$$z_0 = \frac{\widehat{p}_1 - \widehat{p}_2}{\sqrt{\widehat{p}(1 - \widehat{p})(\frac{1}{n_1} + \frac{1}{n_2})}},$$

where $\widehat{p} = (x_1 + x_2)/(n_1 + n_2)$. We reject H_0 at the α level of significance if $|z_0| > z(\alpha/2)$. Since the square of a standard normal variable $N(0, 1)$ has a chi-square distribution χ_1^2, it can be shown after some algebra that the square of the above statistic is the well-known Pearson's chi-square test given in Sect. 4.4.2 later.

The reader should be aware that the whole subject of comparing two probabilities is a diverse one as there are a variety of opinions expressed in the literature on the most appropriate procedure. With the advent of better computing facilities, exact rather than approximate methods are now being recommended. Such methods are discussed in Sect. 4.4.3. These exact and approximate methods have been recently compared by Fagerland et al. (2011), who give some recommendations.

3.3.2 Relative Risk

The difference between two proportions $p_1 - p_2$ is much more critical when the proportions are near 0 (or 1) than near 1/2. For example, the proportion of people experiencing side effects of a particular drug are usually small. If the observed proportions of two drugs are 0.010 and 0.001, then the difference of 0.009 is the same as if the two proportions were 0.510 and 0.501. However, in the first case, the first drug has 10 times as many reactions as the second. A better measure in this case is the relative risk $\phi = p_1/p_2$. Assuming we have observations x_1 and x_2 from two independent Binomial distributions, $\mathrm{Bin}(n_i, p_i)$ ($i = 1, 2$), we would now be interested in obtaining confidence intervals for ϕ. The obvious estimate of ϕ is $\widehat{\phi} = \widehat{p}_1/\widehat{p}_2$, where $\widehat{p}_i = x_i/n_i$. However $\widehat{\phi}$ has a highly skewed sampling distribution unless the samples are very large. To get round this problem we can use a log transformation that works well in normalizing the distribution when the underlying distribution is skewed right, as for example when a binomial distribution has small p_i (Katz et al. 1978). Using a Taylor expansion from (3.31) in the Appendix (Sect. 3.7) below, we find that

$$\mathrm{var}(\log \widehat{P}_i) \approx (1 - p_i)/n_i p_i.$$

Hence for independent samples, we can sum the above expression for $i = 1, 2$, replace each p_i by its estimate, and obtain the following large sample $100(1 - \alpha)\%$ confidence interval for $\log \phi$, namely

$$\log(\widehat{p}_1/\widehat{p}_2) \pm z(\alpha/2)\sqrt{\frac{1 - \widehat{p}_1}{x_1} + \frac{1 - \widehat{p}_2}{x_2}}.$$

Taking exponentials of the above limits give us a confidence interval for ϕ. This interval does not work well for small or moderate sample sizes but we can improve the situation by adding 1/2 to each item giving us

$$\log \frac{(x_1 + 0.5)(n_2 + 0.5)}{(n_1 + 0.5)(x_2 + 0.5)} \pm z(\alpha/2)\sqrt{\frac{1}{x_1 + 0.5} + \frac{1}{x_2 + 0.5} - \frac{1}{n_1 + 0.5} - \frac{1}{n_2 + 0.5}}.$$

When one of the two outcomes has a small probability (p_i or q_i) we work with the ratio of the probabilities for that outcome. For example, if the p_i's are near 1 we can work with the observations $n_i - x_i$, and with q_i instead of p_i. An exact procedure is available in Sect. 3.4.2 below when the probabilities are from the same population.

A number of other methods are available and these are described and compared by Fagerland et al. (2011).

3.3.3 Odds Ratio

One other parameter of interest is the so called "odds," $p/(1 - p)$, and the "odds ratio"

$$\theta = \frac{p_1/(1 - p_1)}{p_2/(1 - p_2)}.$$

When $p_1 = p_2$, $\theta = 1$. Replacing each p_i by $\widehat{p}_i$, we can then estimate θ by $\widehat{\theta} = x_1(n_2 - x_2)/x_2(n_1 - x_1)$. Once again we can use a log transformation, and, from a Taylor expansion (Sect. 3.7.1), we get

$$\mathrm{var}\left[\log\left(\frac{X_i}{n_i - X_i}\right)\right] \approx \left(\frac{1}{n_i p_i} + \frac{1}{n_i q_i}\right)^2 \mathrm{var}(X_i) = \frac{1}{n_i p_i} + \frac{1}{n_i q_i} = \frac{1}{n_i p_i q_i}.$$

As X_1 and X_2 are independent, we an add the variances and then replace parameters by their estimates giving us the following large sample $100(1 - \alpha)\%$ confidence interval for θ, namely (Woolf 1955)

$$\widehat{\theta} \pm z(\alpha/2)\sqrt{\frac{1}{x_1} + \frac{1}{n_1 - x_1} + \frac{1}{x_2} + \frac{1}{n_2 - x_2}}. \tag{3.9}$$

To avoid problems with zero elements one can add 0.5 to each element (Gart 1966) giving us the estimate

$$\widetilde{\theta} = \frac{(x_1 + 0.5)(n_2 - x_2 + 0.5)}{(x_2 + 0.5)(n_1 - x_1 + 0.5)}.$$

The variance expression can also be adjusted by adding 0.5 to each element in the variance. Agresti (1999) discussed these adjustments and provided an alternative adjustment based on the idea of "smoothing toward independence." The odds ratio is considered further in the next chapter when discussing 2×2 contingency tables. Exact confidence intervals are discussed there. Various methods for θ are described and compared by Fagerland et al. (2011).

3.4 Multinomial Distribution

If we can approximate sampling without replacement by sampling with replacement, we can set $r = 1$ above, and the Multi-Hypergeometric distribution (3.1) can be replaced by the Multinomial distribution with probability function

$$f(\mathbf{x}) = \frac{n!}{x_1! x_2! \cdots x_k!} p_1^{x_1} p_2^{x_2} \cdots p_k^{x_k}, \quad \sum_{1=1}^{k} p_i = 1. \tag{3.10}$$

As $\sum_{1=1}^{k} x_i = n$, this form of the multivariate distribution, as with (3.1), is singular and has a singular variance-covariance matrix, $\boldsymbol{\Sigma} = \{\text{cov}(x_i, x_j)\}$. Its one advantage is its symmetric formulation. Setting $r = 1$ in Sect. 3.1, we get $\text{var}(X_i) = n p_i q_i$ and $\text{cov}(X_i, X_j) = -n p_i p_j$. If $x. = \sum_{i=1}^{k-1} x_i$, a nonsingular formulation is

$$g(x_1, \ldots, x_{k-1}) = \frac{n!}{x_1! x_2! \cdots x_{k-1}! (n - x.)!} p_1^{x_1} p_2^{x_2} \cdots p_{k-1}^{x_{k-1}} \left(1 - \sum_{i=1}^{k-1} p_i\right)^{n-x.}, \tag{3.11}$$

which we use below. The Multinomial distribution also arises when we have n fixed Bernoulli trials but with k possible outcomes rather than just two, as with the Binomial distribution.

Arguing as in Sect. 3.1, we can add probabilities together to show that any subset is also Multinomial. In particular, the marginal distribution of X_i is $\text{Bin}(n, p_i)$. For further properties of the Multinomial distribution see Johnson et al. (1997, Chap. 35).

3.4.1 Maximum Likelihood Estimation

Referring to the nonsingular multinomial distribution, the log-likelihood function (excluding constants) is a function of $\mathbf{p} = (p_1, p_2, \ldots, p_{k-1})'$, namely

$$\log L(\mathbf{p}) = \sum_{i=1}^{k-1} x_i \log(p_i) + x_k \log(p_k),$$

where $p_k = 1 - \sum_{i=1}^{k-1} p_i$ and $x_k = n - \sum_{i=1}^{k-1} x_i$. Then

$$\frac{\partial \log L(\mathbf{p})}{\partial p_i} = \frac{x_i}{p_i} - \frac{x_k}{p_k} = 0 \quad \text{for } i = 1, 2, \ldots, k - 1. \tag{3.12}$$

Solving these equations leads to $\widehat{p}_i = x_i \widehat{p}_k / x_k$, and summing $i = 1, 2, \ldots, k$ gives us

$$1 = \sum_{i=1}^{k} \widehat{p}_i = n \frac{\widehat{p}_k}{x_k} \quad \text{so that} \quad \widehat{p}_i = \frac{x_i}{n} \quad \text{for } i = 1, 2, \ldots, k.$$

Replacing observations x_i by random variables X_i, we find that the vector $\widehat{\mathbf{P}} = n^{-1}(X_1, X_2, \ldots, X_{K-1})'$ has mean $\mathbf{p}$ and positive-definite variance-covariance matrix $n^{-1}\boldsymbol{\Sigma}$, where

$$\boldsymbol{\Sigma} = \begin{pmatrix} p_1 q_1 & -p_1 p_2 & -p_1 p_3 & \cdots & -p_1 p_{k-1} \\ -p_2 p_1 & p_2 q_2 & -p_2 p_3 & \cdots & -p_2 p_{k-1} \\ \cdot & \cdot & \cdot & \cdots & \cdot \\ -p_{k-1} p_1 & -p_{k-1} p_2 & -p_{k-1} p_3 & \cdots & p_{k-1} q_{k-1} \end{pmatrix}$$
$$= \text{diag}(\mathbf{p}) - \mathbf{p}\mathbf{p}'.$$

Here $\text{diag}(\mathbf{p})$ is a diagonal matrix with diagonal elements $p_1, \ldots, p_{k-1}$. From Seber (2008, result 15.7), with $0 < p_i < 1$ for all i,

$$\boldsymbol{\Sigma}^{-1} = \text{diag}(\mathbf{p}^{-1}) + p_k^{-1}\mathbf{1}_{k-1}\mathbf{1}_{k-1}',$$

where $\text{diag}(\mathbf{p}^{-1})$ has diagonal elements p_i^{-1}. The inverse of a positive definite matrix is also positive definite (Seber 2008, result 10.27).

To prove that $\widehat{\mathbf{P}}$ maximizes the likelihood, we note first that

$$-\frac{\partial^2 \log L(\mathbf{p})}{\partial p_i \partial p_j} = \delta_{ij}\frac{x_i}{p_i^2} + \frac{x_k}{p_k^2} \quad (i, j = 1, 2, \ldots k - 1), \tag{3.13}$$

where $\delta_{ij} = 1$ if $i = j$ and 0 otherwise. As the matrix

$$-\left\{\frac{\partial^2 \log L(\mathbf{p})}{\partial p_i \partial p_j}\right\}_{\mathbf{p}=\widehat{\mathbf{p}}} = n[\text{diag}(\widehat{\mathbf{p}}^{-1}) + \widehat{p_k}^{-1}\mathbf{1}_{k-1}\mathbf{1}_{k-1}'] = \boldsymbol{\Sigma}^{-1}_{\mathbf{p}=\widehat{\mathbf{p}}}$$

is positive-definite for $0 < \widehat{p_i} < 1$ $(i = 1, 2, \ldots, k - 1)$, $\widehat{\mathbf{p}}$ is a maximum.

We note from (3.13) that the (expected) information matrix is

$$-E\left\{\frac{\partial^2 \log L(\mathbf{p})}{\partial p_i \partial p_j}\right\} = n\boldsymbol{\Sigma}^{-1}, \tag{3.14}$$

and its inverse is the Cramér-Rao lower bound for random vectors (using the Löwner ordering of positive-definite matrices; see Seber 2008, Sect. 10.1). This inverse is also the variance-covariance matrix of $\widehat{\mathbf{P}}$, which implies that $\widehat{P_i}$ is the unbiased estimator of p_i with minimum variance. More generally, $\mathbf{h}'\widehat{\mathbf{P}}$ is the unbiased estimator of $\mathbf{h}'\mathbf{p}$ with minimum variance.

3.4.2 Comparing Two Probabilities from the Same Population

In this section we consider exact and approximate methods of inference. By summing probabilities over categories for the Multinomial distribution, we see that the sum of any of the X_i is $\mathrm{Bin}(n, \sum_i p_i)$. In particular, $X_1 + X_2$ is $\mathrm{Bin}(n, p_1 + p_2)$ and the conditional distribution of X_1 given $X_1 + X_2$ is

$$
\begin{aligned}
\mathrm{pr}(X_1 = x_1, X_2 = {}&x_2 \mid X_1 + X_2 = x_1 + x_2) \\
&= \frac{\mathrm{pr}(X_1 = x_1, X_2 = x_2)}{\mathrm{pr}(X_1 + X_2 = x_1 + x_2)} \\
&= \binom{x_1 + x_2}{x_1} \left(\frac{p_1}{p_1 + p_2}\right)^{x_1} \left(\frac{p_2}{p_1 + p_2}\right)^{x_2},
\end{aligned}
\tag{3.15}
$$

which is $\mathrm{Bin}(m, p)$, where $m = x_1 + x_2$ and $p = p_1/(p_1 + p_2)$. To test the hypothesis $H_0 : p_1 = p_2$ we can carry out an exact test of $p = \frac{1}{2}$ using the above distribution and the method of Sect. 2.3.2. We can also obtain an exact confidence interval for p, say (p_L, p_U), that can be turned into a confidence interval for the relative risk

$$
\frac{p_2}{p_1} = \frac{1}{p} - 1,
$$

namely $(p_U^{-1} - 1, p_L^{-1} - 1)$.

Arguing as in Sect. 3.2, we can also use (3.4) and (3.5) with $r = 1$ to construct an approximate confidence interval for $p_1 - p_2$ and to test $p_1 = p_2$.

3.5 Asymptotic Multivariate Normality

We saw in Sect. 2.1.4 that an estimate of a probability $\widehat{P}_i = X_i/n$ can be treated as a sample mean, and we now extend this to vectors. Let $\mathbf{Y}$ be a k-dimensional random vector that takes the unit Cartesian vector $\mathbf{y}_i$ with probability p_i ($i = 1, 2, \ldots, k$), where $\mathbf{y}_i$ has all zero elements except 1 in the ith position. If there are n trials, $\mathbf{y}_i$ will have a frequency of x_i ($i = 1, 2, \ldots, k$) in the sample, and the joint probability function of the n $\mathbf{y}$-observations is $\prod_{i=1}^{k} p_i^{x_i}$. We then have the random vector

$$
\begin{aligned}
\widehat{\mathbf{P}}_k &= (\widehat{P}_1, \widehat{P}_2, \ldots, \widehat{P}_k)' \\
&= \frac{1}{n}(X_1, X_2, \ldots, X_k)' \\
&= \frac{1}{n}\sum_{i=1}^{k} X_i \mathbf{y}_i \\
&= \overline{\mathbf{Y}}, \quad \text{say.}
\end{aligned}
$$

If $\widehat{\mathbf{P}} = \widehat{\mathbf{P}}_{(k-1)}$, then, by the Multivariate Central Limit theorem, $\sqrt{n}(\widehat{\mathbf{P}} - \mathbf{p})$ is asymptotically distributed as a $(k-1)$-dimensional Multivariate Normal distribution with mean $\mathbf{0}$ and nonsingular variance-covariance matrix Σ. This result also follows from the general result that a maximum likelihood estimator is asymptotically Normal.

Setting $X_k = n - \sum_{i=1}^{k-1} X_i$, we find that

$$n(\widehat{\mathbf{P}} - \mathbf{p})'(p_k^{-1}\mathbf{1}_{k-1}\mathbf{1}_{k-1}')(\widehat{\mathbf{P}} - \mathbf{p}) = \frac{(n - X_k - n(1 - p_k))^2}{np_k}$$

$$= \frac{(X_k - np_k)^2}{np_k},$$

and, from the properties of the Multivariate Normal distribution (cf. Seber 2008, result 20.25)

$$n(\widehat{\mathbf{P}} - \mathbf{p})'\Sigma^{-1}(\widehat{\mathbf{P}} - \mathbf{p}) = \left\{ \sum_{i=1}^{k-1} \frac{(X_i - np_i)^2}{np_i} \right\} + \frac{(X_k - np_k)^2}{np_k}$$

$$= \sum_{i=1}^{k} \frac{(X_i - np_i)^2}{np_i}$$

$$= \sum_{i=1}^{k} \frac{X_i^2}{np_i} - n \qquad (3.16)$$

is asymptotically distributed as a χ_{k-1}^2.

An alternative and shorter derivation of the Eq. (3.16) involves using the singular Multinomial distribution of $\widehat{\mathbf{P}}_{(k)}$, and by augmenting $\mathbf{p}$ to $\mathbf{p}_{(k)} = (\mathbf{p}', p_k)'$ and Σ to $\Sigma_{(k)}$, say (by changing the subscript $k-1$ to k). Then, $\Sigma_{(k)}^{-} = \mathrm{diag}(p_1^{-1}, p_2^{-1}, \ldots, p_k^{-1})$ is a generalized inverse of $\Sigma_{(k)}$ (as $\Sigma_{(k)}\Sigma_{(k)}^{-}\Sigma_k = \Sigma_{(k)}$). We have immediately (Seber 2008, result 20.29, with $\mathbf{A} = \Sigma_{(k)}$)

$$n(\widehat{\mathbf{P}}_{(k)} - \mathbf{p}_{(k)})'\Sigma_{(k)}^{-}(\widehat{\mathbf{P}}_{(k)} - \mathbf{p}_{(k)}) = \sum_{i=1}^{k} \frac{(X_i - np_i)^2}{np_i}$$

is asymptotically χ_{k-1}^2, as $\Sigma_{(k)}\mathbf{1}_k = \mathbf{0}$ (from $\sum_{i=1}^{k} p_i = 1$) implies that $\Sigma_{(k)}$ has rank $k-1$.

We can now use (3.16) for inference. For example, we can test the hypothesis $H_0 : \mathbf{p} = \mathbf{p}_0$ using the large sample test statistic

$$X_0^2 = n(\widehat{\mathbf{P}} - \mathbf{p}_0)'\Sigma_0^{-1}(\widehat{\mathbf{P}} - \mathbf{p}_0), \qquad (3.17)$$

This is (3.16) with $\mathbf{p}$ replace by $\mathbf{p}_0$, and it has approximately the χ^2_{k-1} distribution when H_0 is true. Here $\boldsymbol{\Sigma}_0$ is $\boldsymbol{\Sigma}$ evaluated at $\mathbf{p} = \mathbf{p}_0$. The above statistic X^2_0 is usually referred to as Pearson's goodness-of-fit test.

3.5.1 Simultaneous Confidence Intervals

If $\chi^2_{k-1}(\alpha)$ is the upper α point (the $1 - \alpha$ quantile) of χ^2_{k-1}, we have from (3.16) with large n,

$$
\begin{aligned}
1 - \alpha &\approx \mathrm{pr}\left[n\,(\widehat{\mathbf{P}} - \mathbf{p})' \boldsymbol{\Sigma}^{-1}(\widehat{\mathbf{P}} - \mathbf{p}) \le \chi^2_{k-1}(\alpha) \right] \\
&= \mathrm{pr}[n\,\mathbf{a}'\boldsymbol{\Sigma}^{-1}\mathbf{a} \le b], \quad \text{say,} \\
&= \mathrm{pr}\left[n \sup_{\mathbf{h}:\mathbf{h}\neq 0} \left\{ \frac{(\mathbf{h}'\mathbf{a})^2}{\mathbf{h}'\boldsymbol{\Sigma}\mathbf{h}} \right\} \le b \right], \quad \text{(from Seber 2008, result 12.1(b)),} \\
&= \mathrm{pr}\left[n \frac{(\mathbf{h}'\mathbf{a})^2}{\mathbf{h}'\boldsymbol{\Sigma}\mathbf{h}} \le b, \text{ all } \mathbf{h}\ (\neq \mathbf{0}) \right] \\
&= \mathrm{pr}\left[\frac{|\mathbf{h}'\widehat{\mathbf{P}} - \mathbf{h}'\mathbf{p}|}{(\mathbf{h}'\boldsymbol{\Sigma}\mathbf{h}/n)^{1/2}} \le \sqrt{\chi^2_{k-1}(\alpha)}, \text{ all } \mathbf{h}\ (\neq \mathbf{0}) \right].
\end{aligned}
$$

Replacing random variables by their observed values, we can therefore construct a confidence interval for *any* linear function $\mathbf{h}'\mathbf{p}$, namely

$$
\mathbf{h}'\mathbf{p} \in \mathbf{h}'\widehat{\mathbf{p}} \pm \sqrt{\chi^2_{k-1}(\alpha)} \left(\frac{\mathbf{h}'\boldsymbol{\Sigma}\mathbf{h}}{n} \right)^{1/2}, \tag{3.18}
$$

and the overall probability for the entire class of such intervals is exactly $1 - \alpha$. However the intervals can be very wide as the class of intervals is large. The above algebra is the same as that used to derive Scheffé's simultaneous intervals that arise in regression analysis (cf. Seber and Lee 2003, p. 123).

There are several things we can now do. By choosing $\mathbf{h}$ to have 1 in the ith position and 0 elsewhere we have $\mathbf{h}'\mathbf{p} = p_i$ and the confidence interval becomes

$$
p_i \in \widehat{p}_i \pm \sqrt{\chi^2_{k-1}(\alpha)} \sqrt{\frac{p_i(1 - p_i)}{n}}, \quad (i = 1, 2, \ldots, k - 1) \tag{3.19}
$$

or (p_{iL}, p_{iU}), where p_{iL} and p_{iU} are the roots of the quadratic

$$
(\widehat{p}_i - p_i)^2 = \chi^2_{k-1}(\alpha) \frac{p_i(1 - p_i)}{n}. \tag{3.20}
$$

By setting $\mathbf{h} = -\mathbf{1}_{k-1}$ (since $p_k - 1 = -\sum_{i=1}^{k-1} p_i$) we see that (3.19) also holds for $i = k$.

Another approach is to replace each p_i by its estimate in $\boldsymbol{\Sigma}$ to get $\widehat{\boldsymbol{\Sigma}}$. Equation (3.18) now becomes

$$\mathbf{h}'\mathbf{p} \in \mathbf{h}'\widehat{\mathbf{p}} \pm \sqrt{\chi_{k-1}^2(\alpha)} \left(\frac{\mathbf{h}'\widehat{\boldsymbol{\Sigma}}\mathbf{h}}{n}\right)^{1/2}, \qquad (3.21)$$

where

$$\mathbf{h}'\widehat{\boldsymbol{\Sigma}}\mathbf{h} = \mathbf{h}'\text{diag}(\widehat{\mathbf{p}})\mathbf{h} - \mathbf{h}'\widehat{\mathbf{p}}\widehat{\mathbf{p}}'\mathbf{h} = \sum_{i=1}^{k-1} h_i^2 \widehat{p}_i - \left(\sum_{i=1}^{k-1} h_i \widehat{p}_i\right)^2. \qquad (3.22)$$

Let $l_i = h_i - h_k$, then, since $\sum_{i=1}^{k} p_i = \sum_{i=1}^{k} \widehat{p}_i = 1$ we have that

$$\sum_{i=1}^{k-1} l_i(\widehat{p}_i - p_i) = \sum_{i=1}^{k-1} h_i(\widehat{p}_i - p_i) - h_k(p_k - \widehat{p}_k) = \sum_{i=1}^{k} h_i(\widehat{p}_i - p_i),$$

and it can be shown that

$$\mathbf{l}'\widehat{\boldsymbol{\Sigma}}\mathbf{l} = \sum_{i=1}^{k-1} l_i^2 \widehat{p}_i - \left(\sum_{i=1}^{k-1} l_i \widehat{p}_i\right)^2 = \sum_{i=1}^{k} h_i^2 \widehat{p}_i - \left(\sum_{i=1}^{k} h_i \widehat{p}_i\right)^2. \qquad (3.23)$$

If $\mathbf{h}$, $\widehat{\mathbf{p}}$, and $\mathbf{p}$ are now expanded to k-dimensional vectors $\mathbf{h}_{(k)}$, $\mathbf{p}_{(k)}$, and $\widehat{\mathbf{p}}_{(k)}$, then the set of all $\mathbf{l}'(\widehat{\mathbf{p}} - \mathbf{p})$ is the same as the set of all $\mathbf{h}'_{(k)}(\widehat{\mathbf{p}}_{(k)} - \mathbf{p}_{(k)})$. This means we can use (3.21) with expanded vectors to include p_k provided we replace $k-1$ by k in $\boldsymbol{\Sigma}$.

By choosing $\mathbf{h}_{(k)}$ with 1 in the ith position and zeros elsewhere, the interval for p_i is then

$$p_i \in \widehat{p}_i \pm \sqrt{\chi_{k-1}^2(\alpha)} \sqrt{\frac{\widehat{p}_i(1 - \widehat{p}_i)}{n}}, \quad i = 1, 2, \ldots, k. \qquad (3.24)$$

Again, by choosing $\mathbf{h}_{(k)}$ appropriately, we also get intervals for all pairwise differences $p_i - p_j$, namely (cf. (3.4) with $r = 1$)

$$p_i - p_j \in \widehat{p}_i - \widehat{p}_j \pm \sqrt{\chi_{k-1}^2(\alpha)} \left(\frac{\widehat{p}_i + \widehat{p}_j - (\widehat{p}_i - \widehat{p}_j)^2}{n}\right)^{1/2}. \qquad (3.25)$$

3.5.2 Bonferroni Confidence Intervals

A different set of intervals, usually referred to as Bonferroni confidence intervals, can be obtained using a Bonferroni inequality (see Seber 2008, Sect. 22.2). Let E_i ($i = 1, 2, \ldots, r$) be the event that the ith confidence interval statement is true, that is it contains the true parameter value with probability $1 - \alpha/r$. If $\overline{E}_i$ denotes the complementary event of E_i, then

$$
\delta = \mathrm{pr}\left(\bigcap_{i=1}^{r} E_i\right) = 1 - \mathrm{pr}\left(\overline{\bigcap_{i} E_i}\right) = 1 - \mathrm{pr}\left(\bigcup_{i} \overline{E}_i\right)
$$

$$
\geq 1 - \sum_{i=1}^{r} \mathrm{pr}(\overline{E}_i) = 1 - \frac{1}{r}\sum_{i=1}^{r} \alpha = 1 - \alpha.
$$

Here δ is the probability of getting all the statements correct, which is at least $1 - \alpha$, the "overall" confidence we can attach to the set of r intervals. As noted by Miller (1981, p. 8), the above inequality is surprisingly sharp, providing r is not too large (say, $r \leq 5$) and α is small, say 0.01. For further comments about this problem and which quantity to control, see Hochberg and Tamhane (1987, pp. 9–12). Applying this method to say r linear combinations $\mathbf{h}'\mathbf{p}$ we replace $\sqrt{\chi_{k-1}^2(\alpha)}$ by $z(\alpha/2r)$ in (3.18), with $r = k$ in (3.21), and $r = t$ in (3.24), where $t = \binom{k}{2}$. Goodman (1965) has indicated that the Bonferroni intervals tend to be shorter than those in (3.24) or (3.25).

3.6 Animal Population Applications

This topic is included as it provides good examples of model building using the above models along with Binomial and Poisson models. This is a very big area of ongoing research and we will consider some of the simpler applications only.

3.6.1 Random Distribution of Animals

Suppose we have a closed animal population of unknown size N in an area of known size A. By closed we mean that there is no movement in or out of the population or any birth or death of animals during the study. The population area is split into k areas of size a_i so that $A = \sum_{i=1}^{k} a_i$. If we assume that the animals move randomly and independently, we have N independent multinomial trials and the probability that an animal is in area a_i is $p_i = a_i/A$. If x_i end up in this area, then the joint distribution of the X_i is the Multinomial distribution with probability function

$$\frac{N!}{\prod_{i=1}^{k} x_i!} \prod_{i=1}^{k} p_i^{x_i}. \tag{3.26}$$

Here the p_i are known. If instead we assume that the number X_i in the ith area has a Poisson distribution Poisson(μ_i) with mean μ_i, and the X_i are mutually independent, then, by the additive property of the Poisson distribution, $T = \sum_{i=1}^{k} X_i$ is Poisson(μ), where $\mu = \sum_{i=1}^{k} \mu_i$. The joint distribution of the X_i, given $T = N$, is

$$\prod_{i=1}^{k} \frac{\mu_i^{x_i}}{x_i!} e^{-\mu_i} \Big/ \frac{\mu^N}{N!} e^{-\mu},$$

which the same as (3.10) but with $p_i = \mu_i/\mu$.

3.6.2 Multiple-Recapture Methods

In Sect. 1.3 we considered a model in which animals could be recaptured once. We now extend this to when an animal can be captured more than once. We begin with a closed population of unknown size N. A simple random sample of animals is captured and given an identifying tag or mark before releasing the sample back into the population and allowing the sample to re-disperse. A second random sample is taken; the marked animals are noted and unmarked animals are marked. This process is repeated k times. Let

$$n_i = \text{size of the } i\text{th sample } (i = 1, 2, \ldots, k),$$
$$x_i = \text{number of tagged animals in } n_i,$$
$$u_i = n_i - x_i,$$
$$M_i = \sum_{j=1}^{i-1} u_i \quad (i = 1, 2, \ldots, k+1)$$
$$= \text{total number of marked animals just before the } i\text{th sample is taken.}$$

Since there are no marked animals in the first sample, we have $x_1 = 0$, $M_1 = 0$, $M_2 = u_1 = n_1$, and we define $M_{k+1} (= r$ say) as the total number of marked animals in the population at the end of the experiment, that is the total number of different animals caught throughout the experiment. Assuming fixed sample sizes, simple random sampling implies we can use Hypergeometric distributions and obtain the conditional probabilities

$$\text{pr}(x_i \mid x_{i-1}, \ldots, x_2, \{n_i\}) = \binom{M_i}{x_i} \binom{N - M_i}{u_i} \Big/ \binom{N}{n_i} \quad (i = 2, 3, \ldots, k).$$

Multiplying these probabilities together gives us, after some algebra, the probability function

$$f_1(x_2, x_3, \ldots, m_k \mid \{n_i\}) = \prod_{i=2}^{k} \binom{M_i}{x_i} \binom{N - M_i}{u_i} \Big/ \binom{N}{n_i}$$

$$= \frac{\prod_{i=2}^{k} \binom{M_i}{x_i}}{\prod_{i=1}^{k} u_i!} \cdot \frac{N!}{(N - r)!} \prod_{i=1}^{k} \binom{N}{n_i}^{-1}, \qquad (3.27)$$

since $N - M_i - u_i = N - M_{i+1}$ and $u_1 = n_1$. We wish to estimate the integer N, which we can do by setting the first difference of the log-likelihood equation equal to zero. We define $\nabla g(N) = g(N) - g(N - 1)$ so that $\nabla \log N! = \log N$. Then

$$\nabla \log f_1 = \log N - \log(N - r) - \sum_{i=1}^{k} (\log N - \log(N - n_i)) = 0$$

or

$$1 - \frac{r}{N} = \prod_{i=1}^{k} \left(1 - \frac{n_i}{N}\right). \qquad (3.28)$$

This is a $(k - 1)$th degree polynomial in N and it has a unique root greater than r (Seber 1982, pp. 586–587). If $[\widehat{N}]$ is the integer part of $\widehat{N}$, then it will be within 1 of the maximum-likelihood estimate. We see that we don't need distinguishing marks for the animals as all we need is r, the number of different animals caught.

If the sampling is determined by effort, then the n_i will be random variables and we can construct a different model as follows. Let a_w be the number of animals with a particular recapture history w, where w is a nonempty subset of the integers $\{1, 2, \ldots, k\}$; for example a_{245} is the number of animals caught in the second, fourth, and fifth samples only, and $r = \sum_w a_w$. If the animals act independently, they can be regarded as N independent "trials" from a multinomial experiment. The joint probability function of the set of random variables $\{a_w\}$ is then

$$f_2(\{a_w\}) = \frac{N!}{\prod a_w!(N - r)!} Q^{N-r} \prod_w P_w^{a_w}, \qquad (3.29)$$

where $Q = 1 - \sum_w P_w$.

If we further assume that all individuals have the same probability $p_i \ (= 1 - q_i)$, and for any individuals the events "caught in the ith sample $(i = 1, 2, \ldots, k)$" are independent, then

$$Q = \prod_{i=1}^{k} q_i, \quad P_{245} = q_1 p_2 q_3 p_4 p_5 q_6 \cdots q_k = \frac{p_2 p_4 p_5 Q}{q_2 q_4 q_5} \text{ etc.}$$

so that (3.29) reduces to (Darroch 1958)

$$f_3(\{a_w\}) = \frac{N!}{\prod_w a_w!(N-r)!} \prod_{i=1}^{k} p_i^{n_i} q_i^{N-n_i} \tag{3.30}$$

From the assumptions, the $\{n_i\}$ are independent Binomial variables,

$$f_4(\{n_i\}) = \prod_{i=1}^{k} \binom{N}{n_i} p_i^{n_i} q_i^{N-n_i}$$

and dividing (3.30) by the above expression give us

$$f_5(\{a_w\} \mid \{n_i\}) = \frac{N!}{\prod_w a_w!(N-r)!} \prod_{i=1}^{k} \binom{N}{n_i}^{-1}.$$

We see that we arrive at the same estimate $\widehat{N}$ as, ignoring constants, the likelihood is the same as (3.27).

The above interplay of Multinomial and Hypergeometric models is typical of the more complex models developed for open populations when migration, birth, and death are taking place (cf. Seber 1982, Sect. 13.1.6).

3.7 Appendix: Delta Method

In this section we consider a well-known method for finding large sample variances. The theory is then applied to the Multinomial distribution. We also consider functions of Normal random variables.

3.7.1 General Theory

We consider general ideas only without getting too involved with technical details about limits. Let X be a random variable with mean μ and variance σ_X^2, and let $Y = g(X)$ be a "well-behaved" function of X that has a Taylor expansion

$$g(X) - g(\mu) = (X - \mu)g'(\mu) + \frac{1}{2}(X - \mu)^2 g'(X_0),$$

where X_0 lies between X and μ and $g'(\mu)$ is the derivative of g evaluated at $X = \mu$. Assuming second order terms can be neglected, we have $\mathrm{E}(Y) \approx g(\mu)$ and

$$
\begin{aligned}
\mathrm{var}(Y) &\approx \mathrm{E}[(g(X) - g(\mu))^2] \\
&\approx \mathrm{E}[(X - \mu)^2][g'(\mu)]^2 \\
&= \sigma_X^2 [g'(\mu)]^2.
\end{aligned}
$$

For example, if $g(X) = \log X$ then, for large μ,

$$
\mathrm{var}(\log X) \approx \frac{\sigma_X^2}{\mu^2}. \tag{3.31}
$$

If $\mathbf{X} = (X_1, X_2, \ldots, X_k)'$ is a vector with mean $\boldsymbol{\mu}$, then for suitable g,

$$
Y = g(\mathbf{X}) - g(\boldsymbol{\mu}) \approx \sum_{i=1}^{k} (X_i - \mu_i) g_i'(\boldsymbol{\mu}) + \ldots,
$$

where $g_i'(\boldsymbol{\mu})$ is $\partial g / \partial X_i$ evaluated at $\mathbf{X} = \boldsymbol{\mu}$. If second order terms can be neglected, we have

$$
\begin{aligned}
\mathrm{var}(Y) &\approx \mathrm{E}[(g(\mathbf{X}) - g(\boldsymbol{\mu}))^2] \\
&\approx \mathrm{E}\left[\sum_{i=1}^{k} \sum_{j=1}^{k} (X_i - \mu_i)(X_j - \mu_j) g_i'(\boldsymbol{\mu}) g_j'(\boldsymbol{\mu}) \right] \\
&= \sum_{i=1}^{k} \sum_{j=1}^{k} \mathrm{cov}(X_i, X_j) g_i'(\boldsymbol{\mu}) g_j'(\boldsymbol{\mu}). \tag{3.32}
\end{aligned}
$$

3.7.2 Application to the Multinomial Distribution

Suppose $\mathbf{X}$ has the Multinomial distribution given by (3.10) and

$$
g(\mathbf{X}) = \frac{X_1 X_2 \cdots X_r}{X_{r+1} X_{r+2} \cdots X_s} \quad (s \leq k).
$$

Then, using the above approach with $\mu_i = n p_i$,

$$
\frac{g(\mathbf{X}) - g(\boldsymbol{\mu})}{g(\boldsymbol{\mu})} \approx \sum_{i=1}^{r} \frac{X_i - \mu_i}{\mu_i} - \sum_{i=r+1}^{s} \frac{X_i - \mu_i}{\mu_i},
$$

and it can be shown that (Seber 1982, pp. 8–9)

$$\text{var}[g(\mathbf{X})] \approx \frac{[g(\boldsymbol{\mu})]^2}{n} \left\{ \sum_{i=1}^{s} \frac{1}{p_i} - (s - 2r)^2 \right\}. \tag{3.33}$$

Two cases of interest in this monograph are, $s = 2r = 2$ and $s = 2r = 4$. In the first case $g(\mathbf{X}) = X_1/X_2$ and

$$\text{var}[g(X)] \approx [g(\boldsymbol{\mu})]^2 \left(\frac{1}{\mu_1} + \frac{1}{\mu_2} \right). \tag{3.34}$$

We are particularly interested in $Y = \log g(\mathbf{X})$, so that from (3.31),

$$\text{var}(Y) \approx \frac{\text{var}[g(\mathbf{X})]}{[g(\boldsymbol{\mu})]^2} = \frac{1}{\mu_1} + \frac{1}{\mu_2}. \tag{3.35}$$

If $g(\mathbf{X})$ is a product of two such independent ratios from independent Binomial distributions, then we just add two more terms to $\text{var}(Y)$. We can estimate $\text{var}(Y)$ by replacing each μ_i by X_i in (3.35).

Using similar algebra, we find that

$$\text{var} \left[\log \left(\frac{X_1 X_2}{X_3 X_4} \right) \right] \approx \sum_{i=1}^{4} \frac{1}{\mu_i}. \tag{3.36}$$

3.7.3 Asymptotic Normality

In later chapters we are interested in functions of a maximum likelihood estimator, which we know is asymptotically Normally distributed under fairly general conditions. For example, suppose $\sqrt{n}(\widehat{\boldsymbol{\mu}}_n - \boldsymbol{\mu})$ is asymptotically $N(\mathbf{0}, \boldsymbol{\Sigma}(\boldsymbol{\mu}))$. Then using the delta method above, $\sqrt{n}(g(\widehat{\boldsymbol{\mu}}) - g(\boldsymbol{\mu}))$ is asymptotically distributed as $N(0, \sigma_g^2)$ as $n \to \infty$, where

$$\sigma_g^2 = \left[\left(\frac{\partial g}{\partial \boldsymbol{\mu}} \right) \boldsymbol{\Sigma}(\boldsymbol{\mu}) \left(\frac{\partial g}{\partial \boldsymbol{\mu}} \right)' \right].$$

This result also holds if we replace g by a vector function $\mathbf{g}$ giving us $N(\mathbf{0}, \boldsymbol{\Sigma_g})$.

References

Agresti, A. (1999). On logit confidence intervals for the odds ratio with small samples. *Biometrics*, 55(2), 597–602.

Agresti, A., & Caffo, B. (2000). Simple and effective confidence intervals for proportions and differences of proportions result from adding two successes and two failures. *The American Statistician*, 54(4), 280–288.

Agresti, A., & Min, Y. (2005). Frequentist performance of Bayesian confidence intervals for comparing proportions in 2 × 2 contingency tables. *Biometrics, 61*(2), 515–523.

Andrés, A. M., & Tejedor, I. H. (2002). Comment on "equivalence testing for binomial random variables: Which test to use?" *The American Statistician, 56*(3), 253–254.

Brown, L., & Li, X. (2005). Confidence intervals for two sample binomial distribution. *Journal of Statistical Planning and Inference, 130*, 359–375.

Darroch, J. N. (1958). The multiple-recapture census. I. Estimation of a closed population. *Biometrika, 45*, 343–359.

Fagerland, M. W., Lydersen, S., & Laake, P. (2011). Recommended confidence intervals for two independent binomial proportions. *Statistical Methods in Medical Research*, to appear. doi:10.1177/0962280211415469.

Gart, J. J. (1966). Alternative analyses of contingency tables. *Journal of the Royal Statistical Society, Series B, 28*, 164–179.

Goodman, L. A. (1965). On simultaneous confidence intervals for multinomial proportions. *Technometrics, 7*, 247–254.

Hochberg, Y., & Tamhane, A. C. (1987). *Multiple comparison procedures*. New York: Wiley.

Johnson, N. L., Kotz, S., & Balakrishnan, A. (1997). *Discrete multivariate distributions*. New York: Wiley.

Katz, D., Baptista, J., Azen, S. P., & Pike, M. C. (1978). Obtaining confidence intervals for the risk ratio in Cohort studies. *Biometrics, 34*, 469–474.

Krishnamoorthy, K., & Thomson, J. (2002). Hypothesis testing about proportions in two finite populations. *The American Statistician, 56*(3), 215–222.

Lahiri, S. N., Chatterjee, A., & Maiti, T. (2007). Normal approximation to the hypergeometric distribution in nonstandard cases and a sub-Gaussian Berry-Esseen theorem. *Journal of Statistical Planning and Inference, 137*, 3570–3590.

Mee, R. W. (1984). Confidence bounds for the difference between two probabilities. *Biometrics, 40*, 1175–1176.

Miettinen, 0., & Nurminen, M. (1985). Comparative analysis of two rates. *Statistics in Medicine, 4*, 213–226.

Miller, R. G, Jr. (1981). *Simultaneous statistical inference* (2nd edn.). New York: Springer-Verlag.

Newcombe, R. G. (1998b). Interval estimation for the difference between two independent proportions: Comparison of eleven methods. *Statistics in Medicine, 17*(8), 873–890.

Scott, A. J., & Seber, G. A. F. (1983). Difference of proportions from the same survey. *The American Statistician, 37*(4), 319–320.

Seber, G. A. F. (1982). *The estimation of animal abundance and related parameters* (2nd edn.). London: Griffin. Also reprinted as a paperback in 2002 by Blackburn Press, Caldwell, NJ.

Seber, G. A. F. (2008). *A matrix handbook for statisticians*. New York: Wiley.

Seber, G. A. F., & Lee, A. J. (2003). *Linear regression analysis* (2nd edn.). New York: Wiley.

Wild, C. J., & Seber, G. A. F. (1993). Comparing two proportions from the same survey. *The American Statistician, 47*(3), 178–181. (Correction: 1994, 48(3):269).

Woolf, B. (1955). On estimating the relation between blood group and disease. *Annals of Human Genetics, 19*, 251–253.

Chapter 4
Multivariate Hypothesis Tests

Abstract We establish the asymptotic equivalence of several test procedures for testing hypotheses about the Multinomial distribution, namely the Likehood-ratio, Wald, Score, and Pearson's goodness-of-fit tests. Particular emphasis is given to contingency tables, especially 2×2 tables where exact and approximate test methods are given, including methods for matched pairs.

Keywords Likelihood-ratio test · Wald test · Score test · Pearson's goodness-of-fit test · Deviance · Freedom equation hypothesis specification · Contingency tables · 2×2 tables · Fisher's exact test · Matched pairs · McNemar's test

4.1 Multinomial Test Statistics

In this section we consider three well-known test statistics, namely the likelihood ratio test, the Wald test, and the Score test. We show that all three test statistics are asymptotically equivalent, and the score statistic is the same as Pearson's goodness of fit test statistic.

4.1.1 Likelihood-Ratio Test for $\mathbf{p} = \mathbf{p}_0$

We begin by considering the likelihood-ratio test for testing $H_0 : \mathbf{p} = \mathbf{p}_0$ for the nonsingular multinomial distribution (3.11), where $\mathbf{p}' = (p_1, p_2, \ldots, p_{k-1})$, $\mathbf{p}_0' = (p_{01}, p_{02}, \ldots, p_{0k-1})$, and $p_{0k} = 1 - \sum_{i=1}^{k-1} p_{0i}$. Since $\widehat{\mathbf{P}}$ (with $\widehat{P}_i = X_i/n$) is a consistent estimator of $\mathbf{p}$, it converges in probability to $\mathbf{p}_0$ when H_0 is true. If $L(\mathbf{p})$ is the likelihood function (ignoring constants), $x_k = n - \sum_{i=1}^{k-1} x_i$, and $\widehat{p}_k = x_k/n$, the likelihood ratio is

G. A. F. Seber, *Statistical Models for Proportions and Probabilities*,
SpringerBriefs in Statistics, DOI: 10.1007/978-3-642-39041-8_4,
© The Author(s) 2013

$$\Lambda = \frac{L(\mathbf{p}_0)}{L(\widehat{\mathbf{p}})} = \frac{\prod_{i=1}^{k} p_{0i}^{x_i}}{\prod_{i=1}^{k} \widehat{p}_i^{x_i}},$$

and

$$G^2 = -2 \log \Lambda$$

$$= 2n \sum_{i=1}^{k} \widehat{p}_i \log \left(\frac{\widehat{p}_i}{p_{0i}} \right)$$

$$= 2n \sum_{i=1}^{k} (\widehat{p}_i - p_{0i} + p_{0i}) \log(1 + y_i) \quad \left(y_i = \frac{\widehat{p}_i - p_{0i}}{p_{0i}} \right)$$

where $\log(1 + y_i) = y_i - y_i^2/2 + y_i^3/3 \ldots$ for $|y_i| < 1$ and y_i converges to 0 in probability. Hence

$$G^2 = 2n \sum_{i=1}^{k} (\widehat{p}_i - p_{0i} + p_{0i}) \left[\frac{\widehat{p}_i - p_{0i}}{p_{0i}} - \frac{(\widehat{p}_i - p_{0i})^2}{2p_{0i}^2} + O_p(\widehat{p}_i - p_{0i})^3 \right]$$

$$= 2n \sum_{i=1}^{k} \left\{ (\widehat{p}_i - p_{i0}) + \frac{(\widehat{p}_i - p_{0i})^2}{p_{0i}} - \frac{(\widehat{p}_i - p_{0i})^2}{2p_{0i}} + O_p(\widehat{p}_i - p_{0i})^3 \right\}$$

$$\approx \sum_{i=1}^{k} \frac{(x_i - np_{0i})^2}{np_{0i}} \quad \left(\text{since} \quad \sum_{i=1}^{k} (\widehat{p}_i - p_{i0}) = 0 \right) \tag{4.1}$$

$$= X^2.$$

This is Pearson's statistic X_0^2 of (3.17).

4.1.2 Wald Test

We note that

$$X_0^2 = \sum_{i=1}^{k} \frac{(x_i - np_{0i})^2}{np_{i0}}$$

$$= n \sum_{i=1}^{k} \frac{(\widehat{p}_i - p_{i0})^2}{\widehat{p}_i} \left\{ 1 - \frac{(\widehat{p}_i - p_{i0})}{\widehat{p}_i} \right\}^{-1}$$

$$= n \sum_{i=1}^{k} \frac{(\widehat{p}_i - p_{i0})^2}{\widehat{p}_i} \left\{ 1 - \frac{(\widehat{p}_i - p_{i0})}{\widehat{p}_i} + O_p(\widehat{p}_i - p_{i0})^2 \right\}$$

$$\approx \sum_{i=1}^{k} \frac{(x_i - n\widehat{p}_i)^2}{n\widehat{p}_i}$$

$$= W, \text{ say},$$

which is known as Wald's test statistic of H_0.

4.1.3 Score Test

There is one other test statistic due to Rao (1973) called the score statistic given by

$$S^2 = \left\{ \left(\frac{\partial \log L(\mathbf{p})}{\partial \mathbf{p}} \right)' \mathbf{D}^{-1} \left(\frac{\partial \log L(\mathbf{p})}{\partial \mathbf{p}} \right) \right\}_{\mathbf{p}=\mathbf{p}_0}$$

$$= n^{-1} \mathbf{l}' \mathbf{\Sigma} \mathbf{l},$$

where

$$l_i = h_i - h_k = \frac{x_i}{p_i} - \frac{x_k}{p_k}$$

(by (3.12)) and $\mathbf{D}^{-1}$ is the inverse of expected information matrix, which is $n^{-1}\mathbf{\Sigma}$ (from (3.14)). Using the identity in $\mathbf{h}$ given by (3.23), replacing $\widehat{p}_i$ by p_{0i}, and setting $h_i = x_i/p_{0i}$, gives us

$$S^2 = n^{-1}(\mathbf{l}'\mathbf{\Sigma}\mathbf{l})_{\mathbf{p}=\mathbf{p}_0}$$

$$= n^{-1} \left[\sum_{i=1}^{k} h_i^2 p_{i0} - \left(\sum_{i=1}^{k} h_i p_{0i} \right)^2 \right]$$

$$= n^{-1} \left[\sum_{i=1}^{k} \frac{x_i^2}{p_{0i}^2} p_{0i} - \left(\sum_{i=1}^{k} \frac{x_i}{p_{0i}} p_{0i} \right)^2 \right]$$

$$= \sum_{i=1}^{k} \frac{x_i^2}{n p_{0i}} - n$$

$$= \sum_{i=1}^{k} \frac{(x_i - n p_{0i})^2}{n p_{0i}}.$$

We see from (4.1) that Pearson's statistic X_0^2 is the score statistic S^2 and it is asymptotically equivalent to the likelihood ratio test statistic and the Wald test statistic for $H_0 : \mathbf{p} = \mathbf{p}_0$. Some computer packages compute all three statistics, though the likelihood ratio and the Pearson (score) statistics are preferred.

The asymptotic equivalence of the three statistics for a more general hypothesis was shown by Seber (1966, Chap. 11) to hold for a random sample from any univariate or multivariate distribution satisfying certain reasonable conditions. In Sect. 3.5 we saw that the Multinomial distribution arises from a random sample of discrete vectors so that the asymptotic equivalence holds for general hypotheses involving this distribution. The asymptotic equivalence for a such a hypothesis was also shown to hold for independent observations with different means in nonlinear regression models (and hence in linear models) by Seber and Wild (1989, Sect. 12.4; $\mathbf{D}$ should be $E(\mathbf{D})$). In both of the above references, the Lagrange multiplier test mentioned there is simply another formulation of the score test. To prove the equivalence of the likelihood ratio, Wald, and Score test statistics, the asymptotic normality of the maximum likelihood estimator is combined with some linearization to show that the problem is asymptotically equivalent to testing a linear hypothesis for a Normal linear model. In this case, all three test statistics are equal.

Summing up, we see that X^2 and G^2 take the memorable forms

$$X^2 = \sum \frac{(O - E)^2}{E} \quad \text{and} \quad G^2 = 2 \sum O \log\left(\frac{O}{E}\right),$$

where $O = $ Observed and $E = $ Expected.

4.1.4 Deviance

In preparation for future chapters we introduce the concept of deviance. Let $\ell_S = \log L(\widehat{\mathbf{p}})$ denote the maximum of the log likelihood for the Multinomial distribution. This model is referred to as a *saturated* model as the number of parameters equals the number of observations and we have a perfect fit of data to model. If ℓ_0 is the maximised log likelihood when H_0 is true, then the likelihood-ratio test given in Sect. 4.1.1 is $G^2 = 2(\ell_S - \ell_0)$. This difference is also called the *deviance* and it is used for comparing models in this and the next chapter.

4.2 A More General Hypothesis

In the previous section we used a constraint form for the null hypothesis H_0, namely $\mathbf{A}(\mathbf{p}) = \mathbf{p} - \mathbf{p}_0 = \mathbf{0}$, where more generally $\mathbf{A}$ is a matrix function of $\mathbf{p}$ and not just a linear function. We now consider a different type of hypothesis.

4.2.1 Freedom Equation Specification of H_0

In genetics we are often interested in testing whether probabilities have a specified structure. For example, the Hardy-Weinberg (H-W) law explains the constancy of allele and genotype frequencies through the generations. With just two alleles consisting of dominant A and recessive a with respective probabilities θ and $1 - \theta$, the possible genotypes are AA, aa, and Aa. The H-W hypothesis states that $p_1 = \text{pr}(AA) = \theta^2$, $p_2 = \text{pr}(aa) = (1 - \theta)^2$, and $p_3 = \text{pr}(Aa) = 2\theta(1 - \theta) = 1 - p_1 - p_2$, that is $p_i = p_i(\theta)$, a function of θ. In human blood genetics there are a number of different genetic markers (cf. Greenwood and Seber 1992), the most common being the ABO system with alleles A, B, and O and corresponding probabilities p_A, p_B, and p_O ($= 1 - p_A - p_B$). Assuming the H-W law, the probabilities of the phenotypes A, B, AB, and O are, respectively, $p_1 = \text{pr}(AA \text{ and } AO) = p_A^2 + 2p_A p_O$, $p_2 = \text{pr}(BB \text{ and } BO) = p_B^2 + 2p_B p_O$, $p_3 = \text{pr}(AB) = 2p_A p_B$, and $p_4 = \text{pr}(OO) = p_O^2 = 1 - p_1 - p_2 - p_3$. These probabilities are obtained by expanding $1 = (p_A + p_B + p_O)^2$. In this case $p_i = p_i(\theta)$, where $\theta' = (\theta_1, \theta_2) = (p_A, p_B)$. We see, in general, that we are interested in testing $H_0 : p_i = p_i(\theta)$ for $i = 1, 2, \ldots, k - 1$, where θ ($\in \Theta$) is a q-dimensional vector of unknown parameters. This form of H_0 is sometimes referred to as a freedom equation specification of the hypothesis. Again it can be shown to be asymptotically equivalent to testing a linear hypothesis for a Normal linear model (Seber 1964, Theorem 1; 1967).

4.2.2 General Likelihood-Ratio Test

We begin by making a number of basic assumptions concerning the parameter $\theta_T = (\theta_{T1}, \theta_{T2}, \ldots, \theta_{Tq})'$, the true value of θ, and $\mathbf{p}_T = \mathbf{p}(\theta_T)$ so that Taylor expansions can be made in the neighbourhoods of θ_T and $\mathbf{p}_T$. These are (Agresti 2002, Sect. 14.2):

1. θ_T is not on the boundary of Θ.
2. All $p_{Ti} > 0$.
3. $\partial p_i(\theta)/\partial \theta_j$ is continuous in a neighborhood of θ_T.
4. The matrix $\mathbf{A} = \{\partial p_i(\theta)/\partial \theta_j\}$ has full rank q at θ_T.

The likelihood function is given by

$$L(\mathbf{p}(\theta)) = \prod_{i=1}^{k} p_i(\theta)^{x_i},$$

where $\mathbf{p}' = (p_1, p_2, \ldots p_{k-1})$ and $p_k = 1 - \sum_{i=1}^{k-1} p_i$. Let $\widehat{\theta}$ be the maximum likelihood estimate of θ, obtained by solving $\partial \log L(\mathbf{p}(\theta))/\partial \theta_i = 0$ for $i = 1, 2, \ldots, q$. The likelihood ratio test statistic for H_0 is then

$$\Lambda = \frac{\prod_{i=1}^{k} p_i(\widehat{\boldsymbol{\theta}})^{x_i}}{\prod_{i=1}^{k} \widehat{p}_i^{\,x_i}},$$

where $\widehat{p}_i = x_i/n$. If H_0 is true, then $\widehat{\mathbf{p}}$ and $\mathbf{p}(\widehat{\boldsymbol{\theta}})$ will both tend to $\boldsymbol{\theta}_T$ in probability so that for n sufficiently large, $\widehat{p}_i - p_i(\widehat{\boldsymbol{\theta}})$ will be of small order. Then, using the same argument that led to Eq. (4.1),

$$
\begin{aligned}
G^2 &= -2\log\Lambda \\
&= 2n\sum_{i=1}^{k} \widehat{p}_i \log \frac{\widehat{p}_i}{p_i(\widehat{\boldsymbol{\theta}})} \\
&= 2n\sum_{i=1}^{k} \widehat{p}_i \log\left(1 + \frac{\widehat{p}_i - p_i(\widehat{\boldsymbol{\theta}})}{p_i(\widehat{\boldsymbol{\theta}})}\right) \\
&\approx \sum_{i=1}^{k} \frac{(x_i - np_i(\widehat{\boldsymbol{\theta}}))^2}{np_i(\widehat{\boldsymbol{\theta}})} = X^2.
\end{aligned}
$$

When H_0 is true, the likelihood-ratio test G^2 is asymptotically distributed as Chi-square with $k - 1 - q$ degrees of freedom, the latter being the difference in the number of "free" parameters for the unconstrained model and the model for H_0 (see also Agresti 2002, Chap. 14). Referring above to Sect. 4.1.4, we see that $G^2 = 2(\ell_S - \ell_1)$, where ℓ_1 is the maximum value of the log likelihood under H_0, and G^2 is the deviance for testing H_0.

As with linear models, the validity of the multinomial model under H_0 can be checked by looking at certain residuals. The raw residuals $e_i = x_i - np_i(\widehat{\boldsymbol{\theta}})$ can be scaled in various ways to give a number of other residuals. For example, the $e_i/\sqrt{np_i(\widehat{\boldsymbol{\theta}})}$ are referred to as Pearson residuals and satisfy $\sum_i e_i^2 = X^2$ above. If we divide the Pearson residuals by their estimated standard deviations, we get the so-called adjusted residuals (Haberman 1974, p. 139) and we can treat them as being approximately independently distributed as $N(0, 1)$ under H_0. A general formula for such residuals is given by Seber and Nyangoma (2000, p. 185) as well as more complex residuals available from nonlinear theory called projected residuals. These residuals have smaller bias, where the bias results from so-called intrinsic curvature effects. Certain components of deviance, called deviance residuals, can also be used for diagnostics.

4.3 Contingency Tables

A two-way table of data as set out in Table 4.1 is described as a contingency table. This data set can arise from a number of experimental situations that will be described in detail below.

Table 4.1 Two-way
contingency table

			Categories			Row totals
		1	2	...	J	
	1	x_{11}	x_{12}	...	x_{1J}	r_1
Group	2	x_{21}	x_{22}	...	x_{2J}	r_2
	.	.	.	...	.	.
	I	x_{I1}	x_{I2}	...	x_{IJ}	r_I
	Column totals	c_1	c_2	...	c_J	n

4.3.1 Test for Independence in a Two-Way Table

We now use the above theory to consider a contingency table where data from a single Multinomial distribution are set out in the form of an $I \times J$ table of observations x_{ij} ($i = 1, 2, \ldots, I; j = 1, 2, \ldots, J$), as in Table 4.1, with p_{ij} being the probability of falling in the (i, j)th category or cell and $\sum_{i=1}^{I} \sum_{j=1}^{J} x_{ij} = n$. For example, we might be interested in seeing if "handedness" and "ocular dominance" are related; ocular dominance (eye dominance or eyedness) is the tendency to prefer visual input from one eye to the other. The three row categories ($I = 3$) are "left-handedness," "right-handedness," and "mixed handedness" (ignoring those who are ambidextrous, which is rare), while the three column categories ($J = 3$) are "left-eyed," "right-eyed," and "ambiocular." Our hypothesis of interest is that handedness is independent of ocular dominance, that is, we have row and column independence or $H_0 : p_{ij} = \alpha_i \beta_j$, where $\sum_{i=1}^{I} \alpha_i = 1$ and $\sum_{j=1}^{J} \beta_j = 1$. We therefore have $\mathbf{p} = \mathbf{p}(\boldsymbol{\theta})$, where $\boldsymbol{\theta}' = (\alpha_1, \ldots \alpha_{I-1}, \beta_1, \ldots, \beta_{J-1}) = (\boldsymbol{\alpha}', \boldsymbol{\beta}'), q = I - 1 + J - 1$ and $k = IJ$. We note that

$$\sum_{i=1}^{I} \sum_{j=1}^{J} p_{ij} = \sum_{i=1}^{I} \alpha_i \sum_{j=1}^{J} \beta_j = 1.$$

Also, it is readily shown that H_0 is equivalent to testing $p_{ij} = p_{i.} p_{.j}$, where $p_{i.} = \sum_j p_{ij}$ and $p_{.j} = \sum_i p_{ij}$.

We require the maximum likelihood estimates of the α_i and the β_j. The likelihood function is

$$L(\boldsymbol{\alpha}, \boldsymbol{\beta}) = \prod_{i=1}^{I} \prod_{j=1}^{J} (\alpha_i \beta_j)^{x_{ij}}$$

$$= \prod_{i=1}^{I} \alpha_i^{r_i} \prod_{j=1}^{J} \beta_j^{c_j},$$

where $r_i = \sum_{j=1}^{J} x_{ij}$ (the ith row sum) and $c_j = \sum_{i=1}^{I} x_{ij}$ (the jth column sum). Using Lagrange multipliers λ_1 and λ_2, we need to differentiate

$$\log L(\alpha, \beta) + \lambda_i \left(\sum_{i=1}^{I} \alpha_i - 1 \right) + \lambda_2 \left(\sum_{j=1}^{J} \beta_j - 1 \right)$$

with respect to α_i and β_j. Hence we solve

$$\frac{r_i}{\alpha_i} + \lambda_1 = 0, \quad \sum_{i=1}^{I} \alpha_i = 1 \quad \text{and}$$

$$\frac{c_j}{\beta_j} + \lambda_2 = 0, \quad \sum_{j=1}^{J} \beta_j = 1.$$

We see that $\lambda_1 = \lambda_2 = -\sum_{i=1}^{I} r_i = -\sum_{j=1}^{J} c_j = -n$, and our maximum likelihood estimates are

$$\widehat{\alpha}_i = \frac{r_i}{n} \quad \text{and} \quad \widehat{\beta}_j = \frac{c_j}{n}.$$

Hence

$$p_{ij}(\widehat{\theta}) = \widehat{\alpha}_i \widehat{\beta}_j = \frac{r_i c_j}{n^2},$$

and the statistic for the test of independence is therefore the much used Pearson's Chi-square statistic

$$X^2 = \sum_{i=1}^{I} \sum_{j=1}^{J} \frac{(x_{ij} - np_{ij}(\widehat{\theta}))^2}{np_{ij}(\widehat{\theta})} = \sum_{i=1}^{I} \sum_{j=1}^{J} \frac{(x_{ij} - r_i c_j/n)^2}{r_i c_j/n}, \qquad (4.2)$$

which, under H_0, is approximately distributed as the Chi-square distribution with

$$k - 1 - q = IJ - 1 - (I - 1) - (J - 1) = (I - 1)(J - 1)$$

degrees of freedom. This approximation gets better with increasing np_{ij} and is usually reasonable if the $np_{ij} \geq 5$ for all i and j in 2×2 tables (though better methods are given below), and all $np_{ij} \geq 1$ with no more than 20 % of the cells having $np_{ij} < 5$ for larger tables (Cochran 1954).

Residuals for checking the validity of the model under H_0 are

$$\frac{x_{ij} - \widehat{\mu}_{ij}}{\sqrt{\widehat{\mu}_{ij}(1 - r_i/n)(1 - c_j/n)}},$$

where $\widehat{\mu}_{ij} = r_i c_j/n$. As these are approximately independently distributed as $N(0, 1)$ under H_0, any cell whose residual exceeds 2 or 3 in absolute value indicates lack of fit in that cell. If IJ is at least 20 we can expect at least one cell to have a value exceeding 2 by chance.

4.3.2 Several Multinomial Distributions

Suppose that we have an $I \times J$ contingency table as in Table 4.1, but with a different independent Multinomial distribution for each row. The observation x_{ij}, with corresponding probability p_{ij}, is now the frequency of the jth category in the ith Multinomial distribution ($i = 1, 2, \ldots, I : j = 1, 2, \ldots, J$), and the row totals r_i are now fixed. We also have $\sum_{j=1}^{J} p_{ij} = 1$ for $i = 1, 2, \ldots, I$. Let $\mathbf{p}'_{(i)} = (p_{i1}, p_{i2}, \ldots, p_{iJ})$, then the so-called hypothesis of homogeneity is $H_0 : \mathbf{p}_{(1)} = \mathbf{p}_{(2)} = \cdots = \mathbf{p}_{(I)} (= \mathbf{p}_{(0)}, \text{say})$, where $\mathbf{p}'_{(0)} = (p_{01}, p_{02}, \ldots, p_{0J})$. We shall now derive the likelihood-ratio test for H_0.

For the unconstrained model, the likelihood function is

$$L(\{p_{ij}\}) = \prod_{i=1}^{I} \prod_{j=1}^{J} p_{ij}^{x_{ij}},$$

and treating each Multinomial distribution independently, the maximum likelihood estimate of p_{ij} is $\widehat{p}_{ij} = x_{ij}/r_i$. Under H_0, the likelihood function is now

$$L(\mathbf{p}_{(0)}) = \prod_{j=1}^{J} \left(\prod_{i=1}^{I} p_{0j}^{x_{ij}} \right) = \prod_{j=1}^{J} p_{0j}^{c_j},$$

where $c_j = \sum_{i=1}^{I} x_{ij}$ is the jth column sum. Using a Lagrange multiplier for $\sum_{j=1}^{J} p_{0j} = 1$, we get the maximum likelihood estimate $\widehat{p}_{0j} = c_j/n$, where $n = \sum_{i=1}^{I} r_i$. The likelihood ratio is

$$\Lambda = \frac{\prod_{j=1}^{J} \widehat{p}_{0j}^{c_j}}{\prod_{i=1}^{I} \prod_{j=1}^{J} \widehat{p}_{ij}^{x_{ij}}},$$

and

$$-2 \log \Lambda = \sum_{i=1}^{I} \sum_{j=1}^{J} x_{ij} \log \frac{\widehat{p}_{ij}}{\widehat{p}_{0j}}.$$

When H_0 is true, $\widehat{p}_{ij} (= x_{ij}/r_i)$ and $\widehat{p}_{0j}$ will tend to the same limit. Using the same argument that led to Eq. (4.1),

$$-2 \log \Lambda = 2 \sum_{i=1}^{I} \sum_{j=1}^{J} r_i \widehat{p}_{ij} \log \left(1 + \frac{\widehat{p}_{ij} - \widehat{p}_{0j}}{\widehat{p}_{0j}} \right)$$

$$\approx 2 \sum_{i=1}^{I} \sum_{j=1}^{J} r_i \frac{(\widehat{p}_{ij} - \widehat{p}_{0j})^2}{2\widehat{p}_{0j}}$$

$$= \sum_{i=1}^{I} \sum_{j=1}^{J} \frac{\left(x_{ij} - r_i c_j / n\right)^2}{r_i c_j / n}. \tag{4.3}$$

The number of "free" parameters for the general model is $I(J-1)$ and the number for the hypothesized model is $J-1$, with the difference being $I(J-1) - J - 1 = (I-1)(J-1)$. From likelihood theory, (4.3) is asymptotically Chi-square with $(I-1)(J-1)$ degrees of freedom when H_0 is true. We see that the above statistic is the same as (4.2).

When $J=2$ and we want to test whether the probabilities from I independent Binomial distributions are equal or not, we can use an exact procedure involving the Multi-hypergeometric distribution (Williams 1988). Simultaneous confidence intervals for comparing these probabilities are available using the studentized-range distribution with a score statistic (Agresti et al. 2008). The method can be applied to a variety of measures, including the difference of proportions, odds ratio, and relative risk.

4.4 2 × 2 Contingency Tables

The above tests of independence and homogeneity are large sample tests and may not be appropriate with small samples. Of particular interest are 2×2 tables, where homogeneity now refers to the special case of comparing independent binomial distributions. These tables provide a good starting point for considering some of the difficulties and have generated considerable research and some controversy. One of the reasons for this is that such tables arise from variety of experimental situations, illustrated by the examples below, that can cause confusion. Methods tend to fall into one of three categories: no fixed marginal sums, one fixed marginal sum, and both fixed marginal sums.

4.4.1 Examples

Example 1. Suppose we wish to compare the proportions of males and females over 21 years who smoke. One method is to take a simple random sample of the same size from each of the male and female populations and count the number of smokers in each sample. The model for this is that of comparing two independent Hypergeometric distributions. If the sampling fractions are small, this reduces to comparing two independent Binomial distributions, a special case of Sect. 4.3.2. For this example we have fixed row sums.

Example 2. We consider the same problem as Example 1, but this time we begin by taking a simple random sample of size n from a combined population of males and females of size N and then classify the people selected in terms of two categories—smoking status and gender. The model for the four subpopulations is the Multi-hypergeometric distribution of Sect. 3.1, or, if the sampling fraction n/N is negligible, we have the Multinomial distribution. This time we have random row and column sums. Some would argue that smoking status is the response variable and gender the explanatory variable so that we could treat the numbers of males and females as (conditionally) fixed and analyze the data as in Example 1.

Example 3. We wish to study the effect on aspirin on the survival of stroke victims. A group of stroke victims is randomly assigned to a treatment group or a placebo group and the number of heart attacks (myocardial infarctions) are recorded for each group over several years. This is a prospective sampling design (called a clinical trial or experimental study) and it would generally be treated as comparing two independent Binomial distributions even though the numbers of trials in each group are random (though approximately equal). We could arrange to have equal numbers in each group by having an even number of patients, then putting them in a random order, numbering them and assigning odd numbers to one group and even numbers to the other group.

Example 4. We wish to study the relationship between smoking and lung cancer. Clearly we cannot assign one group to smoking and another group to nonsmoking. Instead we can use a retrospective design and carry out a so-called case-control (observational) study whereby we take a group of lung cancer patients (cases) and compare them with a same-sized group of those who don't have lung cancer (controls). We then look at the past smoking history for each group. Here we can treat the data as two independent populations—cases and controls. The problem is that the proportions estimate the conditional probabilities of smoking for the two groups given lung cancer status when what we really want are the reverse conditional probabilities, namely the conditional probabilities of getting lung cancer given smoking status. As Agresti (2002, p. 42) notes, we cannot estimate the latter without knowing the proportion of the overall population having lung cancer and then use Bayes' theorem. This means that we cannot estimate differences or ratios of probabilities of lung cancer; only for comparisons of the two probabilities of being a smoker. However, McCullagh and Nelder (1983, p. 113)[1] showed that the odds ratio θ of Sect. 3.3.3 can still be estimated and can then be interpreted conditionally in either direction.

Example 5. Sometimes, in an example like 4 above, the controls are matched to the cases using a number of variables such as age, sex, weight, and so on. Members of a matched pair are not independent so that alternative methods of inference are needed. These are discussed in Sect. 4.4.4.

4.4.2 Chi-Square Test

We return to the large sample chi-square tests for independence and homogeneity. After some algebra we find that X^2 of (4.3) for 2 × 2 tables reduces to

---

[1] See also Agresti (2002, pp. 45–46).

$$X^2 = \sum_{i=1}^{2} \sum_{j=1}^{2} \frac{x_{ij}^2}{r_i c_j / n} - n$$

$$= \frac{n(x_{11}x_{22} - x_{12}x_{21})^2}{r_1 r_2 c_1 c_2}.$$

Here X^2 is asymptotically χ_1^2 when H_0 is true. A better test is available if we replace n by $n - 1$ and the expected cell counts are at least 1.[2] If Yates' correction for continuity (Yates 1934) is used, X^2 becomes

$$\frac{n(|x_{11}x_{22} - x_{12}x_{21}| - 0.5n)^2}{r_1 r_2 c_1 c_2}.$$

However, the exact methods given in the next section along with modern software have made this controversial correction of historical interest only and should not be used (Hirji 2006, p. 149; Lydersen et al. 2009, p. 1170).

We note that $H_0 : p_{ij} = p_{i.} p_{.j}$ for independence reduces to $p_{11}p_{22} - p_{12}p_{21} = 0$ or

$$\theta = \frac{p_{11}/p_{12}}{p_{21}/p_{22}} = \frac{p_{11} p_{22}}{p_{12} p_{21}} = 1.$$

The parameter θ is called the "odds ratio" (also called the "cross-products ratio") for this design using a Multinomial distribution, and its maximum-likelihood estimate is $\widehat{\theta} = x_{11}x_{22}/x_{12}x_{21}$. As the distribution of $\widehat{\theta}$ is highly skewed, even for quite large n, it is preferable to use $\log \widehat{\theta}$ as it is more like a Normal random variable $N(\log \theta, \sigma^2)$, where, using a Taylor expansion, σ is estimated by (cf. (3.36))

$$\widehat{\sigma} = \sqrt{\frac{1}{x_{11}} + \frac{1}{x_{12}} + \frac{1}{x_{21}} + \frac{1}{x_{22}}}.$$

An approximate $100(1 - \alpha)\%$ confidence interval for $\log \theta$ is $\log \widehat{\theta} \pm z(\alpha/2)\widehat{\sigma}$ or (a, b), giving the corresponding interval (e^a, e^b) for θ. This so-called Woolf logit interval does not perform well if any count is small. If, however, we replace each x_{ij} by $x_{ij} + 0.5$ in $\widehat{\theta}$ and $\widehat{\sigma}$ (Agresti 2007, p. 31) we obtain Gart's adjusted logit interval that always performs quite well and can be recommended without judging whether some count is "small" or not.

In the case of testing for homogeneity we are comparing two independent Binomial distributions and θ becomes $p_1(1 - p_2)/p_2(1 - p_1)$. We see that $\theta = 1$ if and only if $p_1 = p_2$. As we saw in Sect. 3.3.3, the same confidence interval for θ applies here. Bayesian confidence intervals and their relative performance compared to frequentist intervals are discussed by Agresti and Min (2005) and Proschan and Nason (2009).

---

[2] Richardson (1994) and Campbell (2007).

4.4.3 Fisher's Exact Test

When some of the x_{ij} are small, Fisher in 1934 proposed an exact test[3] of $H_0 : \theta = 1$ that could be applied to either the multinomial model with four cells or to two independent binomial samples. The method consists of considering all the tables that have the same row (r_i) and column (c_j) sums as the observed table and evaluating the probability distribution of x_{11} given H_0 is true. By conditioning on all the row and column sums, one x_{ij} determines the other three, and therefore the whole table. If we have two independent Binomial distributions and $\theta = p_1(1 - p_2)/[p_2(1 - p_1)]$, then it can be shown that

$$\text{pr}(X_{11} = x_{11} \mid X_{11} + X_{21} = c_1)$$

$$= \binom{r_1}{x_{11}}\binom{r_2}{c_1 - x_{11}}\theta^{x_{11}} \Big/ \sum_{u=0}^{c_1}\binom{r_1}{u}\binom{r_2}{c_1 - u}\theta^u.$$

This distribution can be used to obtain "exact" confidence intervals for θ (Troendle and Frank 2001).

When $\theta = 1$, $p_1 = p_2$, and the probability function of X_{11} is the Hypergeometric distribution

$$\text{pr}(x_{11}) = \binom{r_1}{x_{11}}\binom{r_2}{x_{21}} \Big/ \binom{n}{c_1}$$

$$= \frac{r_1! r_2! c_1! c_2!}{x_{11}! x_{12}! x_{21}! x_{22}! n!}. \tag{4.4}$$

If the data come from a Multinomial distribution, then conditional on r_1 (which also fixes r_2), we find that X_{11} and X_{21} have independent Binomial distributions, namely for $i = 1, 2$, $X_{i1} \sim \text{Bin}(r_i, p_{i1}/p_i)$, where $p_i = p_{i1} + p_{i2}$. This leads to the case above when we fix c_1 and c_2, but with the difference that $\theta = p_{11}p_{22}/p_{12}p_{21}$.

In both cases, if the hypothesis is $H_0 : \theta \leq 1$, the alternative hypothesis is $H_a : \theta > 1$, and $\hat{\theta} = x_{11}x_{22}/x_{12}x_{21} > 1$, then we reject H_0 if the probability of getting a value greater than or equal to the observed value of $\hat{\theta}$ is small enough (e.g., less than α). This probability can be computed exactly. If $H_a : \theta \neq 1$, then several two-sided tests are possible (Agresti 2007, p. 93). A common method, already previously mentioned, is to define the two-sided p-value as twice the smallest tail (TST), that is twice the smallest of the one-sided p-values. Agresti makes the important point: "To conduct a test of size 0.05 when one truly believes that the effect has a particular direction, it is safest to conduct the one-sided test at the 0.025 level to guard against criticism." These issues are discussed further by Hirji (2006, pp. 206–210).

The above so-called conditional tests are somewhat conservative, especially when the groups sizes are small or the probabilities are close to 0 or 1. One adjustment that

---

[3] As in previous chapters exact means using the exact values of the underlying probability distribution rather than an approximation.

has support is to use the so-called mid p-value discussed in Sects. 2.3.1 and 2.3.2. As noted there, the mid p-value is half the probability of the observed result plus the probability of more extreme values. The conditional mid-p test is less conservative but does not always preserve the test size. Its performance approximates that of an unconditional test (Lydersen et al. 2009, p. 1168).

One can also invert the exact test to construct exact confidence intervals. Again a mid-p adjustment can be used. Agresti and Min (2001) showed that inverting a two-sided test is usually better than inverting two-one sided tests. Details of the general hypothesis theory are given by Lehmann and Romano (2005, Sects. 4.5 and 4.6). Software is available for such tests, for example, StatXact version 9.[4]

There has been some controversy over the use of the above exact test as it is a *conditional* test, conditional on c_1 and c_2 (Agresti 1992, p. 148). Mehrotra et al. (2003) discussed various unconditional exact tests and provided evidence that an exact test based on the score statistic outperforms Fisher's test, as does Boschloo (1970) test, in which the p-value from Fisher's test is used as the test statistic. Their key message is that care is needed in choosing an unconditional test. Lydersen et al. (2009) recommend the use of unconditional exact tests, which means not conditioning on any marginals that are not fixed by the design. Their paper should be consulted for details.

4.4.4 Correlated Data

The test for homogeneity in a 2×2 table requires that the two Binomial distributions are independent. In many situations this is not the case as the data from the two distributions are correlated. For example, consider the following scenarios with possibly increasing correlation.

Example 6. Suppose we have a sample of people (the target population) exposed to a certain situation (e.g., radiation) and the number possessing a certain attribute, A say, (e.g., cancer) is recorded. We then match up each individual in that group with a similar individual who has not been exposed and record the number with A in the matched sample. This will give us the table of data, Table 4.2 that we return to later. We want to test whether there is a difference between the two groups with respect to attribute A.

Example 7. We want to answer the question of whether members of a couple tend to possess the same characteristic (e.g., tall or not tall). This is an observational study that would be based on a random sample of couples. The two groups, men and women, form matched samples, matched by marriage.

Example 8. One way of getting matched pairs in an experiment is to use a random sample of twins. For example, we may wish to examine the effectiveness of some "treatment" by assigning the treatment randomly to one of the pair and the "control"

---

[4] See http://www.cytel.com/pdfs/SX9_brochure_final-2.pdf.

Table 4.2 Matched pairs

		Treatment 2	
		A present	A absent
Treatment 1	A present	x_{11}	x_{12}
	A absent	x_{21}	x_{22}

to the other member. Here we would expect the degree of correlation between the samples to be greater than in Example 7.

Example 9. The strongest correlation occurs when the same group of people is used for each sample. For instance, given a random sample of people, each person undergoes two tests on separate occasions to see if there is a negative or positive result for each test. Another example is when each person has to respond "Yes" or "No" to each of two questions, as in Sect. 3.2.2.

These examples can be treated the same way as Example 6, that we now return to but with the exposed sample referred to as receiving Treatment 1 and the unexposed sample as receiving Treatment 2 for more generality. From Table 4.2, we see that x_{11} and x_{22} tell us nothing about the difference in the two samples as they yield the same outcome in each case.

However, if the two samples were the same with regard to the presence or absence of A, then we would expect x_{12} and x_{21} to be approximately equal. Our hypothesis of no difference is then $H_0 : p_{12} = p_{21}$. If $x = x_{12} + x_{21}$, then, under H_0, each of the x observations has a probability of 1/2 of contributing to x_{12} and a probability of 1/2 of contributing to x_{21}. Conditioning on x, we can treat X_{12} as being Bin(x, p), and H_0 is now $p = 1/2$. We can therefore carry out an exact test of $p = 1/2$ using the theory of Sect. 2.3.2, and most software packages do this for all x. Under H_0, X_{12} has mean $x/2$ and variance $x/4$ so that we can we can use a Normal approximation for $x > 10$ (though some recommend larger values of x). Hence, when H_0 is true, we have the score test

$$z_0 = \frac{(x_{12} - x/2)}{\sqrt{x/4}} = \frac{x_{12} - x_{21}}{\sqrt{x_{12} + x_{21}}},$$

which is approximately $N(0, 1)$, and z_0^2 is approximately χ_1^2. The latter test is known as McNemar's test (McNemar 1947). If a continuity correction is used, the test becomes

$$z_0^2 = \frac{(|x_{12} - x_{21}| - 1)^2}{x_{12} + x_{21}}.$$

However, such corrections are not necessary when we have an exact test, though asymptotic methods considered below can be useful in a classroom setting.

We saw in the example of Sect. 3.2.2 that if p_1 is the probability of A being present in the target sample and p_2 the same probability for the matched sample, then $\delta = p_1 - p_2 = p_{12} - p_{21}$. We now consider constructing a confidence interval for δ. Our estimator of δ is $\widehat{D} = \widehat{P}_1 - \widehat{P}_2 = (X_{12} - X_{21})/n$ with mean δ and (assuming we have a Multinomial situation or we can ignore the sampling fraction if Hypergeometric) with variance (see (3.6))

$$\sigma^2(\widehat{D}) = \frac{1}{n}\left\{p_{12} + p_{21} - (p_{12} - p_{21})^2\right\}.$$

The Wald large sample confidence interval for δ is then

$$\widehat{d} \pm z(\alpha/2)\,\widehat{\sigma}(\widehat{D}),$$

where $d = (x_{12} - x_{21})/n$ and $\widehat{\sigma}(\widehat{D})$ is $\sigma(\widehat{D})$ with each p_{ij} replaced by x_{ij}/n. This interval does not perform well with small samples.

Under $H_0 : \delta = 0$, and setting $p_{12} - p_{21} = 0$ we obtain an alternative variance estimate

$$\widehat{\sigma}_0^2(\widehat{D}) = \frac{\widehat{p}_{12} + \widehat{p}_{21}}{n} = \frac{x_{12} + x_{21}}{n^2}.$$

The score test statistic then becomes

$$z_0 = \frac{d}{\widehat{\sigma}_0(\widehat{D})} = \frac{x_{12} - x_{21}}{\sqrt{x_{12} + x_{21}}},$$

which is McNemar's test once again.

A confidence interval for δ can be obtained by inverting the score test for testing $H_0 : \delta = d$ using a method analogous to Mee's method described in Sect. 3.3.1. The method requires the maximum likelihood estimation of p_{12} and p_{21} subject to $p_{12} - p_{21} = d$ (Tango 1998).

A natural extension to the matched pairs problem in which two probabilities are compared is to measure further binary characteristics so that we end up with multivariate binary data. Agresti and Klingenberg (2005), in discussing this problem, give the example where one group are given a drug while the other group are given a placebo. The percentages of the groups having various side effects are then compared. We then have two vectors to compare, one for the drug and the other for the placebo.

References

Agresti, A. (1992). A survey of exact inference for contingency tables. *Statistical Science, 7*(1), 131–153.

Agresti, A. (2002). *Categorical data analysis.* New York: Wiley.

Agresti, A. (2007). *An introduction to categorical data analysis.* Hoboken, NJ: Wiley-Interscience.

Agresti, A., & Klingenberg, B. (2005). Multivariate tests comparing binomial probabilities, with application to safety studies for drugs. *Applied Statistics, 54*(4), 691–706.

Agresti, A., & Min, Y. (2001). On small-sample confidence intervals for parameters in discrete distributions. *Biometrics, 57*, 963–971.

Agresti, A., & Min, Y. (2005). Frequentist performance of Bayesian confidence intervals for comparing proportions in 2 × 2 contingency tables. *Biometrics, 61*, 515–523.

Agresti, A., Bini, M., Bertaccini, B., & Ryu, E. (2008). Simultaneous confidence intervals for comparing binomial parameters. *Biometrics, 64*(4), 1270–1275.

Boschloo, R. D. (1970). Raised conditional level of significance for the 2×2 table when testing the equality of probabilities. *Statistica Neerlandica, 24*, 1–35.

Campbell, I. (2007). Chi-squared and Fisher-Irwin tests of two-by-two tables with small sample recommendations. *Statistics in Medicine, 26*, 3661–3675.

Cochran, W. G. (1954). Some methods for strengthening the common chisquare tests. *Biometrics, 10*, 427–451.

Greenwood, S. R., & Seber, G. A. F. (1992). Estimating blood phenotype probabilities and their products. *Biometrics, 48*, 143–154.

Haberman, S. J. (1974). *The analysis of frequency data.* Chicago: University of Chicago Press.

Hirji, K. F. (2006). *Exact analysis of discrete data.* Boca Raton: Chapman & Hall.

Lehmann, E. L., & Romano, J. P. (2005). *Testing statistical hypotheses* (3rd edn.). New York: Springer.

Lydersen, S., Fagerland, M. W., & Laake, P. (2009). Tutorial in biostatistics: Recommended tests for association in 2×2 tables. *Statistics in Medicine, 28*, 1159–1175.

McCullagh, P. & Nelder, J. A. (1983). *Generalized linear models.* London: Chapman & Hall. (Second edition, 1989.)

McNemar, Q. (1947). Note on the sampling error of the difference between correlated proportions or percentages. *Psychometrika, 12*, 153–157.

Mehrotra, D. V., Chan, I. S. F., & Berger, R. L. (2003). A cautionary note on exact unconditional inference for a difference between two independent binomial proportions. *Biometrics, 59*, 441–450.

Proschan, M. A., & Nason, M. (2009). Conditioning in 2×2 tables. *Biometrics, 65*(1), 316–322.

Rao, C. R. (1973). *Linear statistical inference and its applications.* New York: Wiley.

Richardson, J. T. E. (1994). The analysis of 2×1 and 2×2 contingency tables: An historical review. *Statistical Methods in Medical Research, 3*(2), 107–133.

Seber, G. A. F. (1964). The linear hypothesis and large sample theory. *Annals of Mathematical Statistics, 35*(2), 773–779.

Seber, G. A. F. (1966). *The linear hypothesis: A general theory.* London: Griffin. (Revised in 1980.)

Seber, G. A. F. (1967). Asymptotic linearisation of non-linear hypotheses. *Sankhyā, The Indian Journal of Statistics, Series A, 29*(2), 183–190.

Seber, G. A. F., & Nyangoma, S. O. (2000). Residuals for multinomial models. *Biometrika, 87*(1), 183–191.

Seber, G. A. F., & Wild, C. J. (1989). *Nonlinear regression.* New York: Wiley. Also printed as a paperback, Wiley.

Tango, T. (1998). Equivalence test and confidence interval for the difference in proportions for the paired-sample design. *Statistics in Medicine, 17*, 891–908.

Troendle, J. F., & Frank, J. (2001). Unbiased confidence intervals for the odds ratio of two independent binomial samples with application to case-control data. *Biometrics, 57*(2), 484–489.

Williams, D. A. (1988). Tests for differences between several small proportions. *Journal of the Royal Statistical Society, Series C (Applied Statistics), 37*(3), 421–434.

Yates, F. (1934). Contingency table involving small numbers and the χ^2 test. *Supplement to the Journal of the Royal Statistical Society, 1*(2), 217–235.

Chapter 5
Logarithmic Models

Abstract In this chapter we consider log-linear and logistic models for handling contingency tables, multinomial distributions, and binomial data. The role of the deviance in hypothesis testing is discussed. The log-linear model is applied to an epidemiological problem involving the merging of incomplete lists.

Keywords Log-linear models · Contingency tables · Logistic models · Multiple-recapture · Epidemiology

5.1 Log-Linear Models

There is a large class of models called generalized linear models where an appropriate transformation transforms a parameter such that we have a linear regression involving explanatory variable(s), which may be continuous or categorical.[1] This is a big subject and the aim of this chapter is to provide some general background only for just two logarithmic transformations to give some idea of the models used.

5.1.1 Contingency Tables

There is a close parallel between the analysis of logarithmic models for contingency tables and analysis of variance models (see Seber and Lee 2003, Chap. 8) that we now demonstrate. Given an $I \times J$ contingency table from a Multinomial distribution with $\mu_{ij} = np_{ij}$, we can consider a model of the form

$$\log \mu_{ij} = \lambda + \lambda_i^{(1)} + \lambda_j^{(2)} + \lambda_{ij}^{(12)}, \tag{5.1}$$

---

[1] See, for example, McCullagh and Nelder (1989) and Hardin and Hilbe (2007).

G. A. F. Seber, *Statistical Models for Proportions and Probabilities*,
SpringerBriefs in Statistics, DOI: 10.1007/978-3-642-39041-8_5,
© The Author(s) 2013

where the $\lambda_{ij}^{(12)}$ are usually referred to as first-order interactions. On the left-hand side we have IJ parameters and on the right-hand side $1+I+J+IJ$ parameters so that we need to apply some constraints to the above parameters for identifiability, as we would in a two-way analysis of variance. The simplest constraints are $\lambda_I^{(1)} = \lambda_J^{(2)} = 0$, $\lambda_{Ij}^{(12)} = 0$ ($j = 1, 2, \ldots, J$), and $\lambda_{iJ}^{(12)} = 0$ ($i = 1, 2, \ldots, I - 1$), giving a total of IJ estimable parameters, one for each observation. We now find that the hypothesis of independence between rows and columns, that is the independence of the two categorical variables defining the table, is true when the $\lambda_{ij}^{(12)}$ are all zero, as then

$$\mu_{ij} = e^{\lambda} e^{\lambda_i^{(1)}} e^{\lambda_j^{(2)}} = a_i b_j,$$

say. In fact it can be shown that

$$\log \mu_{ij} = \lambda + \lambda_i^{(1)} + \lambda_j^{(2)}$$

if and only if $p_{ij} = p_{i.}p_{.j}$ for all i and j, where $p_{i.} = \sum_j p_{ij}$ etc. (Christensen 1997, pp. 49–50). This result is readily demonstrated for a 2×2 table, as the log odds is given by (cf. (5.1))

$$\log \theta = \log \left\{ \frac{p_{11}p_{22}}{p_{12}p_{21}} \right\} = \log \left\{ \frac{\mu_{11}\mu_{22}}{\mu_{12}\mu_{21}} \right\} = \lambda_{11}^{(12)} + \lambda_{22}^{(12)} - \lambda_{12}^{(12)} - \lambda_{21}^{(12)},$$

a contrast in the interactions. This contrast is zero (i.e. $\theta = 1$) if and only if all the $\lambda_{ij}^{(12)}$ are zero. For an $I \times J$ table, each odds ratio corresponds to an interaction contrast and the contrast is zero if the ratio is 1. If all these contrasts are zero, then the interactions are all zero. A similar result holds if we now have independent Multinomial distributions. The interactions are zero if and only if the Multinomial distributions have the same probabilities, that is, $p_{1j} = p_{2j} = \cdots = p_{Ij}$, $j = 1, 2, \ldots, J$.

The above method can be applied to higher-way tables. For example, given a three-way contingency table with three categorical variables, observed frequencies x_{ijk}, and $\mu_{ijk} = \mathrm{E}(X_{ijk}) = np_{ijk}$, where $\sum_i \sum_j \sum_k p_{ijk} = 1$, we can fit the model

$$\log \mu_{ijk} = \lambda + \lambda_i^{(1)} + \lambda_j^{(2)} + \lambda_k^{(3)} + \lambda_{ij}^{(12)} + \lambda_{jk}^{(23)} + \lambda_{ik}^{(13)} + \lambda_{ijk}^{(123)}, \qquad (5.2)$$

where the $\lambda_{ij}^{(12)}, \lambda_{jk}^{(23)}$, and $\lambda_{ik}^{(13)}$ are first-order interactions, and the $\lambda_{ijk}^{(123)}$ are second-order interactions. This mimics the higher-way layouts in analysis of variance. In fitting such models we are interested in testing whether some of the parameters are zero, for example $\lambda_{ij}^{(12)} = 0$ for all i, j in the two-way layout modelled by (5.1). In the three-way layout modelled by (5.2) we might start with testing $\lambda_{ijk}^{(123)} = 0$ for all i, j, k, and, if that hypothesis is not rejected, we set all the $\lambda_{ijk}^{(123)}$ equal to zero, and then test whether all or some of the first order interactions are zero for our new model. With all interactions zero we end up with the model

$$\log \mu_{ijk} = \lambda + \lambda_i^{(1)} + \lambda_j^{(2)} + \lambda_k^{(3)},$$

which implies that all three categorical variables defining the table are independent. This is true if and only if all possible odds ratios are equal to 1 (Christensen 1997, pp. 38–40). Successively reducing the number of parameters in a model provides a "nested" approach to hypothesis testing where each model in the sequence is a special case of the previous one. We discuss this process in the next section.

5.1.2 Test Statistics

Suppose we wish to test a hypothesis H_0 for an $I \times J \times K$ three-way table, and the maximum likelihood estimate of $\mu_{ijk} = np_{ijk}$, given H_0 is true, is $\widehat{\mu}_{ijk} = n\widehat{p}_{ijk}$. Then, as we found for a two-way table in the previous chapter, we can use two test statistics, the likelihood ratio and the Pearson goodness-of fit statistics. The Pearson statistic is

$$X^2 = \sum_{i=1}^{I} \sum_{j=1}^{J} \sum_{k=1}^{K} \frac{(x_{ijk} - \widehat{\mu}_{ijk})^2}{\widehat{\mu}_{ijk}}, \qquad (5.3)$$

which is distributed as χ_k^2 when H_0 is true. Here k is $IJK - f$, where f is the number of independent ("free") model parameters. To obtain the likelihood-ratio test statistic we require minus twice the difference in the maximum values of the log likelihoods for the hypothesized model and the underlying model, namely

$$2(\ell_S - \ell_0) = 2 \sum_{i=1}^{I} \sum_{j=1}^{J} \sum_{k=1}^{K} x_{ijk} \log \left(\frac{x_{ijk}}{\widehat{\mu}_{ijk}} \right), \qquad (5.4)$$

where ℓ_0 is the maximum log likelihood under H_0 and ℓ_S is the same for the saturated model (Sect. 4.1.4). We note that (5.3) is a quadratic approximation to (5.4), which can be proved using the method of Sect. 4.1.1.

In the case of nested hypotheses in the three-way table, we might proceed as follows and consider testing H_1 : *all second order interactions zero* using the deviance $2(\ell_S - \ell_1)$. If this is not rejected we can test H_{12} : *all first order interactions zero* assuming H_1 is true using the difference

$$2(\ell_1 - \ell_{12}) = \sum_{i=1}^{I} \sum_{j=1}^{J} \sum_{k=1}^{K} x_{ijk} \log(\widehat{\mu}_{ijk}^{(1)} / \widehat{\mu}_{ijk}^{(12)}) \qquad (5.5)$$

$$= 2(\ell_S - \ell_{12}) - 2(\ell_S - \ell_1),$$

which, from (5.4), is the difference of the two deviances, one for each model. Here $\widehat{\mu}_{ijk}^{(1)}$ and $\widehat{\mu}_{ijk}^{(12)}$ are the estimates under H_1 and H_{12}, respectively. The deviances play the same role as sums of squares in analysis of variance models. We note that each deviance is asymptotically chi-square when its hypothesized model is true. Also, $2(\ell_1 - \ell_{12})$ is asymptotically statistically independent of $2(\ell_S - \ell_1)$, a result that is true quite generally (Seber 1967). From the general theory of log-linear models (e.g., Agresti 2002, p. 364) and a certain orthogonality property that we also meet in analysis of variance models, we have that (5.5) is equal to

$$\sum_{i=1}^{I}\sum_{j=1}^{J}\sum_{k=1}^{K}\widehat{\mu}_{ijk}^{(1)}\log(\widehat{\mu}_{ijk}^{(1)}/\widehat{\mu}_{ijk}^{(12)}),$$

which leads to the quadratic approximation, Pearson's statistic,

$$X^2 = \sum_{i=1}^{I}\sum_{j=1}^{J}\sum_{k=1}^{K}(\widehat{\mu}_{ijk}^{(1)} - \widehat{\mu}_{ijk}^{(12)})^2/\widehat{\mu}_{ijk}^{(12)}.$$

While the difference of the two deviances in (5.5) leads to a likelihood-ratio test, the same does not hold for their quadratic approximations that give us the Pearson statistics. For further details concerning two-way and multiway-way tables see Agresti (2002), Fienberg (1980), and Bishop et al. (1975). Log-linear models can also be expanded to include explanatory quantitative variables as with analysis of covariance in linear models (Christensen 1997, Chap. 7).

5.1.3 Application of Log Linear Models to Epidemiology

A common epidemiological problem is that of estimating the number n of people with a certain disease (e.g., diabetes) from several incomplete lists (e.g., doctors' records, hospital records, pharmaceutical records, society memberships, and population surveys). Each list "captures" a certain proportion of the n people and will miss some, so we then end up with several incomplete lists. This approach is called the multiple-recapture or multiple-records method and has been applied to wide variety of populations, for example drug addicts and homeless people to name two.

To estimate the number n with a certain disease we need to be able to estimate the number of people with the disease not on any list. We do this by constructing a model for the lists and then use the model to estimate those not captured at all. If there are K lists, we then have 2^K capture histories with each history consisting of a K-dimensional vector consisting of 0's and 1's, where 1 denotes captured and 0 denotes not captured. For example, if $K = 3$, an individual with the history $(1, 0, 1)$ represents being on lists 1 and 3 but not on list 2. In general, the data give us an incomplete 2^K contingency table, incomplete as we don't know the number not on any list with capture history $(0, 0, \ldots, 0)$. The model we endeavor to fit is the log linear model of Sect. 5.1.1 to the multi-dimensional contingency table.

We now demonstrate the method for the simplest case of $K = 2$. The capture histories now take the form $(1, 1)$, $(0, 1)$, $(1, 0)$, and $(0, 0)$ with corresponding frequencies n_{11}, n_{01}, n_{10}, and n_{00}, where n_{00} is unknown. Using the model (5.1) with $\mu_{ij} = np_{ij}$, where p_{ij} is the probability of having capture history (i, j), and applying identifiability restrictions to reduce the number of parameters, we can express the model in the form

$$\log \mu_{11} = \lambda + \lambda_1^{(1)} + \lambda_1^{(2)} + \lambda_{11}^{(12)}$$
$$\log \mu_{01} = \lambda - \lambda_1^{(1)} + \lambda_1^{(2)} - \lambda_{11}^{(12)}$$
$$\log \mu_{10} = \lambda + \lambda_1^{(1)} - \lambda_1^{(2)} - \lambda_{11}^{(12)}$$
$$\log \mu_{00} = \lambda - \lambda_1^{(1)} - \lambda_1^{(2)} + \lambda_{11}^{(12)}.$$

As we have four parameters but only three observations n_{ij}, we have to make at least one assumption. If the lists are independent, then $\mu_{11} = np_1 p_2$, $\mu_{01} = nq_1 p_2$, etc., where p_i $(= 1 - q_i)$ is the probability of being on the ith list. The interaction term $\lambda_{11}^{(12)}$ is now zero, thus reducing the number parameters by 1.

An alternative parametrization that gives clearer meaning to the parameters is as follows:

$$\log \mu_{11} = \lambda$$
$$\log \mu_{01} = \lambda + \lambda_1^{(1)}$$
$$\log \mu_{10} = \lambda + \lambda_1^{(2)}$$
$$\log \mu_{00} = \lambda + \lambda_1^{(1)} + \lambda_1^{(2)} + \lambda_{11}^{(12)},$$

where $\lambda_{11}^{(12)} = \log[p_{00} p_{11}/(p_{01} p_{10})]$, the log odds ratio. With independence, $\lambda_{11}^{(12)} = 0$, $\lambda = \log(np_1 p_2)$, $\lambda_1^{(1)} = \log(q_1/p_1)$, and $\lambda_1^{(2)} = \log(q_2/p_2)$. As we have three observations n_{11}, n_{01}, and n_{10}, we can estimate λ, $\lambda_1^{(1)}$, and $\lambda_1^{(2)}$, and hence estimate n (and n_{00}) using a GLIM package. Both parametrizations lead to the same estimate of n.

The assumption of independence will hold if one of the two lists is a random sample from the general population. In this case estimates can be found more directly (cf. Sect. 1.3 with N instead of n) without recourse to a log-linear model, which is used here for demonstration purposes. For K lists we can use a general log linear model, but we need to make at least one assumption as in the two-list model, the usual assumption being the highest order interaction is zero. We can then test whether other interactions can be set equal to zero, and once we arrive at the final model we can use it to estimate n. The idea of using a log-linear model for an incomplete 2^K contingency table was introduced by Fienberg (1980) and developed further by Cormack (1989) in the capture-recapture context. Some epidemiological background is given in two papers written by an international working group (IWG 1995a,b). Problems arising from list dependence and heterogeneity are discussed in the first paper. An additional complication arises when errors are made in forming the lists (e.g., mistyping a name) that requires a modification of the multinomial model (cf. Seber et al. 2000; Lee et al. 2001). The method adapts a technique from ecology (cf. Seber and Felton 1981) that uses the notion of double tagging where the information on a person (e.g., name, age, sex, blood pressure, blood glucose level etc.) is split into two parts representing two "tags." A mistake in transmitting a part of this information is equivalent to a either a one-tag or two-tag loss.

5.2 Logistic Models

In a Binomial experiment with n trials, one might want to model a relationship between p, the probability of "success," and an explanatory variable x. One simple model is $p(x) = \beta_0 + \beta_1 x$, but this has the problem that $p(x)$, or its estimate, may not lie in [0, 1]. We have seen above that we can define $\mu(x) = np(x)$ and then use the log transformation to get, for example, $\log \mu(x) = \alpha + \beta_1 x$, say, where $\alpha = \log n + \beta_0$. Other transformations, however, have been used such as probit$[p(x)]$, logit $[p(x)]$, and $-\log[1 - p(x)]$.

As the relationship between $p(x)$ and x is usually nonlinear, a useful generalised linear model is the so-called logistic model

$$p(x) = \frac{\exp(\beta_0 + \beta_1 x)}{1 + \exp(\beta_0 + \beta_1 x)},$$

so that now $0 \leq p(x) \leq 1$. Taking the log odds,

$$g(x) = \text{logit}\,[p(x)] = \log\left[\frac{p(x)}{1 - p(x)}\right] = \beta_0 + \beta_1 x, \tag{5.6}$$

so that

$$\frac{p(x)}{1 - p(x)} = e^{\beta_0}(e^{\beta_1})^x,$$

which leads to a multiplicative interpretation of the odds.

Agresti (2007, pp. 219–220) described a useful connection between the log-linear and logistic models. Referring to Eq. (5.2), suppose variable 2, say Y, is binary taking values 1 and 2 and we treat it as a response variable, with variables 1 and 3 treated as explanatory (independent) variables at levels i and k respectively. Then, if $\lambda_{ijk}^{(123)} = 0$ for all i, j, k in (5.2), Agresti showed that given $X = i$ and $Z = k$

$$\text{logit}[\,\text{pr}(Y = 1)] = \log\left[\frac{\text{pr}(Y = 1)}{1 - \text{pr}(Y = 1)}\right]$$

$$= \log\left[\frac{\text{pr}(Y = 1 \mid X = i, Z = k)}{\text{pr}(Y = 2 \mid X = i, Z = k)}\right]$$

$$= \log\left(\frac{\mu_{i1k}}{\mu_{i2k}}\right) = \log\mu_{i1k} - \log\mu_{i2k}$$

$$= \alpha + \beta_i^{(1)} + \beta_k^{(3)},$$

where α etc. are linear combinations of the λ parameters. We see then that the loglinear model with its second interaction zero leads to an additive logistic regression model. This type of connection can be utilised to examine interaction effects in log-linear models. We now consider some applications of logistic models.

5.2.1 Independent Binomial Distributions

We first consider k independent Binomial variables Y_i. Suppose when $x = x_i$, Y_i has a Binomial distribution with parameters n_i and $p(x_i)$, where (5.6) holds for each x_i. Then the log likelihood function for the parameters β_0 and β_1 is

$$\log[L(\beta_0, \beta_1)] = \sum_{i=1}^{k}\{y_i \log p(x_i) + (n_i - y_i)\log(1 - p(x_i))\}.$$

Differentiating with respect to β_0 and β_1 leads to the maximum likelihood estimates $\widehat{\beta}_0$ and $\widehat{\beta}_1$, the fitted line $\widehat{\beta}_0 + \widehat{\beta}_1 x$, and estimate

$$\widetilde{p}(x_i) = \frac{\exp(\widehat{\beta}_0 + \widehat{\beta}_1 x_i)}{1 + \exp(\widehat{\beta}_0 + \widehat{\beta}_1 x_i)}.$$

To test H_0 that the logit regression model holds, we can use the likelihood-ratio test statistic (and the deviance)

$$2(\ell_S - \ell_0) = 2\left\{\sum_{i=1}^{k} y_i \log\left(\frac{\widehat{p}_i}{\widetilde{p}(x_i)}\right) + \sum_{i=1}^{k}(n_i - y_i)\log\left(\frac{1 - \widehat{p}_i}{1 - \widetilde{p}(x_i)}\right)\right\}, \quad (5.7)$$

where $\widehat{p}_i = x_i/n_i$ for the saturated model. As there are k p_is and two regression parameters, we find that for large n_i the above deviance is approximately distributed as χ_{k-2}^2 when H_0 is true. Setting $\widetilde{p}_i = \widetilde{p}(x_i)$, the Pearson statistic is

$$X^2 = \sum_{i=1}^{k} \left[\frac{(y_i - n_i \widetilde{p}_i)^2}{n_i \widetilde{p}_i} + \frac{[(n_i - y_i) - n_i(1 - \widetilde{p}_i)]^2}{n_i(1 - \widetilde{p}_i)} \right]$$

$$= \sum_{i=1}^{k} \left[\frac{(y_i - n_i \widetilde{p}_i)^2}{n_i \widetilde{p}_i} + \frac{(y_i - n_i \widetilde{p}_i)^2}{n_i(1 - \widetilde{p}_i)} \right]$$

$$= \sum_{i=1}^{k} \frac{(y_i - n_i \widetilde{p}_i)^2}{n_i \widetilde{p}_i(1 - \widetilde{p}_i)}.$$

The above two test statistics are preferred to the Wald test statistic. For testing $H_1 : \beta_1 = 0$ or constructing a confidence interval for β_1, one can use the usual Wald statistic $\widehat{\beta}_1/SE$, where SE is the estimate of the standard deviation (so-called standard error) of $\widehat{\beta}_1$. However, although the Wald statistic takes a simple form, it is generally preferable to use the likelihood test to test H_1 and then invert this test to obtain a confidence interval for β_1.

Software for carrying out the computations uses an iterative re-weighted least squares estimation because the Binomial variables having different variances. If the n_i are small, then both statistics are not appropriate and this problem is discussed by Hosmer and Lemeshow (2000, Chap. 5). A common situation is when each $n_i = 1$ so that we have one observation for each of k different Binomial trials, and $y_i = 1$ or $y_i = 0$. We then see that $\ell_S = 0$. Ryan (2009, pp. 315–319) has a good exposition of this case. Various other measures of testing goodness-of-fit are available including exact inference (Ryan 2009, Sects. 9.6–9.8; see also Agresti 2002, Sect. 6.7). If possible, the data can be grouped; for example, if x is the age of an individual then ages can be grouped into specific age groups. Various grouping strategies are available (e.g., Hosmer and Lemeshow 2000, Sect. 5.2). Logistic regression can also be applied to qualitative explanatory variables using a model such as logit $(p_i) = \beta_0 + \beta_i$, resembling a one-way analysis of variance model. Another formulation may use dummy variables in the regression model.

All the methods used for analyzing linear models can be applied here, for example, we can include further explanatory variables, construct confidence intervals, and look at diagnostic methods such as residual analysis, model fitting, model selection, and stepwise methods (Hosmer and Lemeshow 2000, Chaps. 4 and 5; Agresti 2007, Chap. 6; Ryan 2009, Chap. 9). Various residuals based on scaled versions of the difference $y_i - n\widetilde{p}_i$ can be used to examine how well a model fits the data. For example, referring to samples from the above binomial models, the Pearson residual takes the form

$$e_i = \frac{y_i - n\widetilde{p}_i}{\sqrt{n_i \widetilde{p}_i(1 - \widetilde{p}_i)}},$$

where $\sum_i e_i^2 = X^2$. The standardised residual is obtained by dividing $y_i - n\widetilde{p}_i$ by its standard error. An alternative residual is the deviance residual, which is the square root of (5.6) with the same sign as the raw residual $y_i - n\widetilde{p}_i$.

One interesting example of fitting two explanatory variables occurs in epidemiology where exposure to a risk factor is recorded as being present or absent and we wish to adjust for a continuous variable $x_2 = z$, say (e.g., age). A possible model is then

$$\text{logit } p(x_1, z) = \beta_0 + \beta_1 x_1 + \beta_2 z,$$

where x_1 takes values 0 or 1 reflecting exposure or not, and z is age. As in linear regression, we may get so-called "confounding" between the two variables so that our model becomes

$$g(x_1, z) = \text{logit } p(x_1, z) = \beta_0 + \beta_1 x_1 + \beta_2 z + \beta_{12} x_1 z,$$

and we have an "interaction" term β_{12}. When $\beta_{12} = 0$ we get parallel lines

$$g(1, z) = \beta_0 + \beta_1 + \beta_2 z \quad \text{and}$$
$$g(0, z) = \beta_0 + \beta_2 z.$$

The above theory with its focus on binary outcomes can also be applied to more than two outcomes, thus leading to logistic modeling of the Multinomial distribution.

5.2.2 Logistic Multinomial Regression Model

We assume we have a nominal response random variable Y that falls in category i with probability p_i ($i = 1, 2, \ldots k$), where $\sum_i^k p_i = 1$. We first pair each category with a baseline category such as the last category k. If there were only two categories i and k so that $p_k = 1 - p_i$, we could work with

$$\log\left(\frac{p_i}{p_k}\right), \quad i = 1, 2, \ldots, k - 1.$$

which would be the log odds for p_i, that is logit p_i. Suppose we have a single explanatory variable x, then we could consider the model

$$g_i(x) = \log\left(\frac{p_i}{p_k}\right) = \alpha_i + \beta_i x, \quad i = 1, 2, \ldots k - 1.$$

We see that the above equation implies for any arbitrary categories r and s that

$$\log\left(\frac{p_r}{p_s}\right) = \log\left(\frac{p_r/p_k}{p_s/p_k}\right) = \log\left(\frac{p_r}{p_k}\right) - \log\left(\frac{p_s}{p_k}\right)$$
$$= (\alpha_r + \beta_r x) - (\alpha_s + \beta_s x)$$
$$= (\alpha_r - \alpha_s) + (\beta_r - \beta_s)x,$$

which is again a straight line. When $k = 2$, we get an ordinary logistic regression for a binary response. When $k = 3$,

$$\frac{p_i(x)}{p_3(x)} = e^{g_i(x)} \quad \text{for } i = 1, 2, 3,$$

where $g_3(x) = 0$, and $\sum_{i=1}^{3} p_i(x) = 1$. Then solving the above equations we get

$$p_3(x) = \frac{1}{1 + \sum_{i=1}^{2} e^{g(x_i)}}$$

and hence

$$p_i(x) = \frac{e^{g_i(x)}}{\sum_{i=1}^{3} e^{g_i(x)}}, \quad i = 1, 2, 3.$$

Suppose that our response variable falls in category i with frequency y_i ($i = 1, 2, \ldots, k$) so that the y_i have a multinomial distribution (cf. Eq. (3.10)), and we assume the straight line regression model above. Then, dropping constants, the likelihood function for the parameters $\boldsymbol{\alpha} = (\alpha_1, \alpha_2, \ldots \alpha_k)'$ and $\boldsymbol{\beta} = (\beta_1, \beta_2, \ldots, \beta_k)'$ is

$$L(\boldsymbol{\alpha}, \boldsymbol{\beta}) = \sum_{i=1}^{k} y_i \log[p(x_i)],$$

where

$$p_i(x) = \frac{e^{g_i(x)}}{\sum_{i=1}^{k} g_i(x)} \quad \text{and} \quad g_k(x) = 0.$$

Software is available for finding the maximum likelihood estimates of the regression parameters. A global test of fit for the regression model can be carried out using the either Pearson's goodness of fit test statistic or the deviance statistic, where the maximum likelihood fitted cell counts are compared with the observed cell counts y_i. As with the binary case $k = 2$ described in the previous section, the above regression model can be extended to include further explanatory variables as well as the diagnostic and model selection techniques generally associated with linear models (e.g., Agresti 2002, pp. 219–229: Christensen 1997, Chap. 6). Computer packages, as well as computing estimates and goodness-of-fit tests, also compute standard errors of various estimated parameters, and produce test statistics for various hypotheses. Agresti (2002), in his Appendix A, has collected together references to software, with an emphasis on SAS, for analyzing categorial data.[2] Ryan (2009, Sect. 9.18) also describes some of the common software packages for logistic regression. Some packages include the Wald and score tests as well as the likelihood ratio test.

---

[2] For additional information, related to the 3rd (2012) edition of his book, on other packages see http://www.stat.ufl.edu/~aa/cda2/cda.html.

Logistic regression can also be used for handling other situations such as, for example, matched studies (Agresti 2002, Chap. 10; 2007, Sect. 8.2; Hosmer and Lemeshow 2000, Chap. 7) and ordinal data (Agresti 2002, Chap. 7 and 2010; Hosmer and Lemeshow 2000, Sect. 8.2).

References

Agresti, A. (2002). *Categorical data analysis*. New York: Wiley.

Agresti, A. (2007). *An introduction to categorical data analysis*. Hoboken, NJ: Wiley-Interscience.

Agresti, A. (2010). *Analysis of ordinal categorical data* (2nd edn.). New York: Wiley.

Bishop, Y. M. M., Fienberg, S. E. & Holland P. W. (with the collaboration of R. J. Light and F. Mosteller). (1975). *Discrete multivariate analysis: Theory and practice*. Cambridge, MA: MIT Press (available electronically).

Christensen, R. (1997). *Log-linear models and logistic regression* (2nd edn.). New York: Springer.

Cormack, R. M. (1989). Log-linear models for capture-recapture. *Biometrics, 45*, 395–413.

Fienberg, S. E. (1980). *The analysis of cross-classified categorical data* (2nd edn.). Cambridge, MA: MIT Press (available electronically).

Hardin, J. W., & Hilbe, J. M. (2007). *Generalized linear models and extensions* (2nd edn.). College Station, TX: Stata Press.

Hosmer, D. W., & Lemeshow, S. (2000). *Applied logistic regression* (2nd edn.). New York: Wiley.

IWG (International Working Group for Disease Monitoring and Forecasting). (1995a). Capture-recapture and multiple-record systems estimation I: History and theoretical development. *American Journal of Epidemiology, 142*(10), 1047–1058.

IWG (International Working Group for Disease Monitoring and Forecasting) (1995b). Capture-recapture and multiple-record systems estimation II: Applications in human diseases. *American Journal of Epidemiology, 142*(10), 1059–1068.

Lee, A. J., Holden, J., Huakau, J., & Seber, G. A. F. (2001). Capture-recapture, epidemiology, and list mismatches: Several lists. *Biometrics, 57*, 707–713.

McCullagh, P., & Nelder, J. A. (1989). *Generalized linear models* (2nd edn.). New York: Chapman and Hall/CRC.

Ryan, T. P. (2009). *Modern regression methods* (2nd edn.). New York: Wiley.

Seber, G. A. F. (1967). Asymptotic linearisation of non-linear hypotheses. *Sankhyā, The Indian Journal of Statistics, Series A, 29*(2), 183–190.

Seber, G. A. F., & Felton, R. (1981). Tag loss and the Petersen mark-recapture experiment. *Biometrika, 48*, 211–219.

Seber, G. A. F., Huakau, J., & Simmons, D. (2000). Capture-recapture, epidemiology, and list mismatches: Two lists. *Biometrics, 56*, 1227–1232.

Seber, G. A. F., & Lee, A. J. (2003). *Linear regression analysis* (2nd edn.). New York: Wiley.